Ecologically Unequal Exchange

R. Scott Frey • Paul K. Gellert
Harry F. Dahms
Editors

Ecologically Unequal Exchange

Environmental Injustice in Comparative and Historical Perspective

palgrave
macmillan

Editors
R. Scott Frey
University of Tennessee
Knoxville, TN, USA

Paul K. Gellert
University of Tennessee
Knoxville, TN, USA

Harry F. Dahms
University of Tennessee
Knoxville, TN, USA

ISBN 978-3-030-07843-0 ISBN 978-3-319-89740-0 (eBook)
https://doi.org/10.1007/978-3-319-89740-0

Softcover re-print of the Hardcover 1st edition 2019

Cover illustration: bCracker/getty images

Printed on acid-free paper

This Palgrave Macmillan imprint is published by the registered company Springer International Publishing AG part of Springer Nature.
The registered company address is: Gewerbestrasse 11, 6330 Cham, Switzerland

Acknowledgments

We would like to thank the authors, reviewers, and our editor at Palgrave, Rachael Ballard, who made this volume possible. Special thanks are due to the University of Illinois Press for giving permission to reproduce a portion of Stephen Bunker's foundational text *Underdeveloping the Amazon* as the first chapter of this volume and to his widow, Dena Wortzel, for her support of the project. We would also like to thank the co-sponsors of our conference on ecologically unequal exchange at the University of Tennessee, including the Center for the Study of Social Justice, the Department of Sociology, the Haines-Morris Grant Program of the College of Arts & Sciences, the Howard H. Baker Center for Public Policy, Africana Studies, Global Studies, Latin American & Caribbean Studies, Asian Studies, and the Department of Anthropology. We give special thanks to Department of Sociology staff and graduate students who helped make the conference possible. And we thank Joel Crombez for compiling the index.

Contents

Notes on Contributors

Daniel Auerbach is a PhD candidate in Sociology at the University of Utah. His primary research interest is the political economy of global environmental change.

Stephen G. Bunker (1944–2005) was Professor of Sociology at the University of Wisconsin, Madison. His research focused on peasant economies in Uganda, Guatemala, Peru, and Brazil. His last works focused on the organization of natural resource extraction, transport, and trade as crucial to the rise of hegemonic powers (with Paul S. Ciccantell) and on how pre-Conquest Andean mountain inhabitants understood and built a system of irrigation in their rugged landscape.

Paul S. Ciccantell is Professor of Sociology at Western Michigan University. His research examines socioeconomic change over the long term, the evolution of global industries, and the socioeconomic and environmental impacts of global industries, focusing particularly on raw materials extraction and processing and transport industries. He is the coauthor of two books with Stephen Bunker, *Globalization and the Race for Resources* and *East Asia and the Global Economy* published by Johns Hopkins University Press. He is writing a book on the coal industry in western Canada.

David Ciplet is Assistant Professor of Environmental Studies at the University of Colorado, Boulder. He is lead author of *Power in a Warming World: The New Global Politics of Climate Change and the Remaking of Environmental Inequality*

(2015) and has published articles in journals such as *Global Environmental Change*, *Global Governance*, and *Global Environmental Politics*.

Brett Clark is Associate Professor of Sociology and Sustainability Studies at the University of Utah. His research focuses on the political economy of global environmental change and the philosophy, history, and sociology of science. He is the coauthor of *The Tragedy of the Commodity: Oceans, Fisheries, and Aquaculture*.

Rebecca Clausen is Chair and Associate Professor of Sociology and Human Services at Fort Lewis College. She is also an affiliate faculty member in the Environmental Studies Program. Her research interests include the social drivers of environmental change, the political economy of global food systems, and marine fishery degradation. She is the coauthor of *The Tragedy of the Commodity: Oceans, Fisheries, and Aquaculture*.

Carolyn Coburn is a PhD candidate in Sociology at the State University of New York at Stony Brook. Her research interests include development, political economy, environmental sociology, and quantitative methods.

Harry F. Dahms is Professor of Sociology, co-director of the Center for the Study of Social Justice, and co-chair of the Committee on Social Theory at the University ofTennessee, Knoxville. He has been the editor ofCurrent Perspectives in Social Theory since 2008 and is the director of the International Social Theory Consortium. His recent work has been in the area of critical theory, and he is in the process of completing a book for Routledge, entitled *Modern Society as Artifice: Critical Theory and the Logic of Capital*.

Mariko Frame is an adjunct at Westminster College's Global Studies, Politics and Justice Program. She is an international political economist who studies the global capitalist economy, development, and the environment. She has lived, researched, and taught throughout Asia and Africa, where the sharp realities of mal-development have driven a research agenda focused on imperialism, inequality, and environmental justice.

R. Scott Frey is Professor of Sociology and co-director of the Center for the Study of Social Justice at the University of Tennessee, Knoxville, and senior associate researcher at the Laboratory for Comparative Social Research at the National Research University Higher School of Economics, Moscow, Russian Federation. His research interests center on the transfer of the core's hazards to the peripheral zones of the world-system, the history and legacy of Agent Orange in Vietnam, and patterns of ecologically unequal exchange in the US Great

Plains. He is working on a book for Routledge, entitled *Globalization, Environmental Health, and Social Justice.*

Paul K. Gellert is Associate Professor of Sociology at the University of Tennessee, Knoxville, and affiliated fellow at the Royal Netherlands Institute of Southeast Asian and Caribbean Studies (KITLV) in Leiden, the Netherlands. His publications on resource nationalism (with B. Kaup), forestry and law, palm oil, and coal (with P. Ciccantell) have appeared in outlets such as *International Journal of Comparative Sociology*, *The Journal of Asian Studies*, and *The Energy and Society Handbook (Oxford)*. He is writing a book on the critical historical sociology of extractive commodities in Indonesia.

Stefano B. Longo is Associate Professor of Sociology at North Carolina State University. His research examines the relationships between human and ecological systems. He is the coauthor of *The Tragedy of the Commodity: Oceans, Fisheries, and Aquaculture*.

Laura McKinney is Assistant Professor of Sociology at Tulane University. Her research focuses on ecology and society, broadly defined, and employs an interdisciplinary approach to nature—society interactions to examine global and local sustainability. Recent publications appear in *Social Forces, Social Problems*, and *Social Science Research*, among other outlets. Current projects fall at the intersection of international development, gender, and the environment, and analyze the prevailing social forces that connect gender to the environment and their collective influence on national and global sustainability.

Jacqueline Patterson is director of the NAACP's Environmental and Climate Justice Program. She co-founded Women of Color United and has served as a senior women's rights policy analyst for ActionAid, where she helps integrate a women's rights lens for food rights, macroeconomics, and climate change into the organization's work.

J. Timmons Roberts is the Ittleson Professor of Environmental Studies and Sociology at Brown University and a non-resident senior fellow at the Brookings Institution. He is author or coauthor of over 80 articles and 10 books, including *Power in a Warming World* and *Fragmented Continent*, both from MIT Press. He serves on the National Research Council's Board on Environmental Change and Society.

John M. Shandra is Associate Professor of Sociology in the Department of Sociology at the State University of New York at Stony Brook. His research interests include environmental sociology, globalization, and statistics.

Jackie Smith is Professor of Sociology at the University of Pittsburgh and editor of an open access journal, *Journal of World-Systems Research*. In addition to publishing numerous books and articles on transnational activism, Smith is co-founder of the International Network of Scholar-Activists and works with technology justice organization, May First/People Link. She coordinates the Pittsburgh Human Rights City Alliance and serves on the National Steering Committee of the US Human Rights Cities Alliance.

Jamie M. Sommer is a PhD candidate in Sociology at the State University of New York at Stony Brook. Her research interests include environmental sociology, global political economy, and development.

Shellen Xiao Wu is Associate Professor of History at the University of Tennessee, Knoxville. Her first book, *Empires of Coal: Fueling China's Entry into the Modern World Order, 1860–1920* (Stanford University Press, 2015), is part of the Weatherhead East Asian Publication series. She has published articles in *The American Historical Review*, *International History Review*, *Journal of Chinese History*, and other leading journals in history and Asian studies. She is working on a second book on frontiers and geopolitical discourse in twentieth-century China.

List of Figures

List of Tables

1

Introduction: Ecologically Unequal Exchange in Comparative and Historical Perspective

R. Scott Frey, Paul K. Gellert, and Harry F. Dahms

At a time of increased societal urgency surrounding ecological crises from depleted fisheries (Longo, Clausen, and Clark 2015) to mineral extraction (Bunker and Ciccantell 2005) and potential pathways toward environmental justice (Martinez-Alier et al. 2016; Smith, Plummer, and Hughes 2016), this collection of papers re-examines ecologically unequal exchange (EUE) in historical and comparative perspective. The theory of EUE, grounded in Wallerstein's (1974–2011) world-systems perspective and the work of Amin (1976), Bunker (1985), and Emmanuel (1972), posits that core or northern consumption and capital accumulation are based on peripheral or southern environmental degradation and extraction. In other words, structures of social and environmental inequality between the Global North and Global South are founded in the extraction of materials from, as well as the displacement of hazardous production processes and wastes to, the Global

R. S. Frey (✉) • P. K. Gellert • H. F. Dahms
University of Tennessee, Knoxville, TN, USA
e-mail: rfrey2@utk.edu; pgellert@utk.edu; hdahms@utk.edu

R. S. Frey et al. (eds.), *Ecologically Unequal Exchange*,
https://doi.org/10.1007/978-3-319-89740-0_1

South (Frey, Gellert, and Dahms 2017; Hornborg and Martinez-Alier 2016; Jorgenson 2016a, 2016b; Jorgenson and Clark 2009a). These unequal relations underscore a large ecological debt owed to the periphery by the core countries; this debt is a key source for many of the previous and current environmental distribution conflicts that have taken place and continue to take place throughout the world-system (Hornborg and Martinez-Alier 2016; Martinez-Alier et al. 2016).

This volume consists of ten chapters based on papers presented at the conference on *Ecologically Unequal Exchange: Environmental Injustice in Historical and Comparative Perspective* held on the campus of the University of Tennessee, Knoxville, on October 15–16, 2015. The conference is part of an ongoing effort by the Center for the Study of Social Justice, housed in the Department of Sociology at the University of Tennessee, to study issues of social justice broadly defined. Additional papers presented at the conference were published in a special issue of the *Journal of World-Systems Research* (Frey et al. 2017).

EUE is an important theory for understanding the uneven socionatural contours of global development and it has fostered research demonstrating that the structure of international trade contributes to environmental degradation in the periphery (see, e.g., Clark and Foster 2009; Frey 2015; Jorgenson 2016b; Jorgenson and Clark 2009b). Various scholars have commented on the "under-utilization" of EUE (Roberts and Parks 2007:195), but there has been a flurry of renewed interest as witnessed by several recent conferences and the publication of several special journal issues devoted to EUE (Frey et al. 2017; Hornborg and Martinez-Alier 2016; Jorgenson and Clark 2009a) and recent contributions by Foster and Holleman (2014), Hornborg (1998, 2009, 2011, 2015), and Jorgenson (2016a, 2016b). In principle, it should be possible to integrate a range of theorists from the ecological economics of Nicholas Georgescu-Roegen (1971) and Howard T. Odum (1971; see also Foster and Holleman 2014) to the promulgation of a theory of EUE by Stephen G. Bunker (1984, 1985, 2005) and from Wackernagel and Rees's (1996) ecological footprint to the ecosocialism of Foster (2002; see also White, Gareau, and Rudy 2017) and the material flow analysis of Marina Fischer-Kowalski (1998; Fischer-Kowalski and Huttler 1999) into a "single theoretical framework clarifying how societal relations of exchange and the material dimensions of production are intertwined," as Alf Hornborg (2009:250) argues.

It is not an easy task to formulate such a single framework, however. As summarized by Oalu (2016:448), the theory has been dubbed "vague and roundabout" (Foster and Holleman 2014:210) and considered at risk of "remaining *scholastic*" (Hornborg 2011:109–114). This lack of specificity in the concept or, more ambitiously, the theory of EUE contributes to the perpetuation of ecological modernization theory's five "interconnected illusions" culminating in the idea that "sustainable development" is possible and not an oxymoron (Hornborg 2009:256). On the other hand, EUE theory can also be considered as *too* specific because it relies on a narrow economistic view of the structure of the capitalist world-economy as hierarchical. Or, at least the methods used to analyze it are economistic in their emphasis on trade data.

In this volume, we try to push scholarship on EUE forward as a useful framework for research by paying attention to the theoretical debates, empirical analyses, and implications for praxis. The authors of the chapters contained in this volume do not all agree with one another—and neither did all of the participants at the conference that formed the basis for this book—yet we believe that presenting them together offers the reader the chance to reflect, adjudicate, and evaluate these contributions. We hope this volume expands critical discussion of EUE.

Book Overview

This volume is organized into three distinct sections: theoretical foundations and critical reflections on EUE; empirical research on economic development, mining, deforestation, fisheries, and the like from the perspective of EUE; and current responses to the adverse socioecological consequences associated with EUE. The first section consists of four chapters examining the theoretical foundations of EUE. Chapter 2 is a condensed reprint of the key theoretical arguments of Stephen G. Bunker's (1985) *Underdeveloping the Amazon: Extraction, Unequal Exchange, and the Failure of the Modern State*. This work is an often cited classic in the field. Yet many have not taken the time to actually read it. By including it here, we hope to re-kindle attention to the formative contributions Bunker made to our understanding of EUE three decades ago. In this book, Bunker argues that the socioecological consequences of extractive economies are quite different from those of production economies. He presents a model of EUE that is a synthesis of various theories of development and underdevelopment, focusing on how local modes of extraction are organized in response to world-system demands. The model is grounded in a case study of the extractive export economies characterizing the Brazilian Amazon Basin during the past 500 years.

Paul Ciccantell in Chapter 3 updates Bunker's original conception by bringing the EUE literature into dialogue with the raw materialism conception that he and Bunker developed (Bunker and Ciccantell 2005). The model focuses attention on "the raw materials-based industries and linked transport systems that are used to solve the most fundamental challenge to rapid economic growth: how to acquire growing volumes of raw materials at lower costs and in greater and more secure volumes than other competing economies." Ciccantell argues that many economies have ascended by using strategies for stealing raw material peripheries from established economies. He goes on to observe that "The current historical juncture in China's economic ascent and in the coal industry creates an opportunity for integrating the insights of ecologically unequal exchange and raw materialism to understand the multidimensional causes and consequences of global inequalities over the very long term."

In Chapter 4, Mariko Frame observes that EUE relations, and the related phenomenon of ecological imperialism, underlie the deep inequalities of the world-system and the exploitation of peripheral countries by core countries. But she notes that semi-peripheral economies are increasingly engaging in economic activities in peripheral countries as they attempt to develop that are as exploitative as those between core and periphery. Frame illustrates her argument by examining land grabbing in Cambodia by various semi-peripheral countries in the Asia region, including Malaysia, Thailand, and Vietnam. Even as these countries are engaging in ecological imperialism with Cambodia and other peripheral countries, they remain subordinate to the core economies and experience adverse socioecological consequences of EUE relations with their core counterparts. Frame concludes that greater theoretical clarity is needed regarding the role played by semi-peripheral countries in EUE relations in the world-system.

Paul K. Gellert in Chapter 5 rounds out the section by returning our attention to Bunker's work, including his posthumously published *The Snake with Golden Braids* (Bunker 2006). He examines the questions of whether and how the seemingly disparate perspectives following Bunker's EUE—specifically Foster's metabolic rift (1999, 2000; Foster, Clark, and York 2010) and Moore's (2000, 2011, 2015; Patel and Moore 2017) world-ecology—are interrelated and in fact complement one another. All three address the unjust manner in which dominant actors in the capitalist world-system simultaneously exploit labor and non-human or biophysical nature while undermining sustainability. Gellert highlights these fundamental agreements regarding EUE across the three perspectives. Then, he unpacks the real distinctions among them regarding capitalism as causing degradation, nature's ontology, epistemology and dialectical analysis, and possible futures that might overturn the current unsustainable situation.

The second section of the book consists of four empirical studies examining EUE in comparative and historical context. Each chapter also stretches the theorization of EUE in fascinating ways. Laura McKinney's Chapter 6 is a cross-national examination of the links between the environment and trajectories of economic development. Drawing on physical science and thermodynamic principles, she presents cross-national data indicating that it is the *liquidation* of resources that stifles economic growth in the periphery, not the abundance of resources identified as the so-called *resource curse*. The structure of the world-system and EUE relationships between core and periphery foster resource appropriation and liquidation that stifle the economic development of the peripheral regions of the world-system. In sum, EUEs and the associated environmental losses in poor nations are driving the wedge that creates unequal economic development.

Jamie Sommer, John Shandra, and Carolyn Coburn in Chapter 7 present cross-national data showing how the flow of mining exports from the periphery to the core affects deforestation in the periphery. Using ordinary least squares regression for a sample of low and middle income nations, they find little support for the expected positive relationship between mining exports and deforestation. They refine their analysis by examining how repressive countries foster EUE in the mining sector by providing a “good business climate” including environmental law exemptions and tax holidays. Their quantitative data indicate that mining export flows from the peripheral countries increase forest loss more in repressive than in democratic countries.

Brett Clark, Stefano Longo, Rebecca Clausen, and Daniel Auerbach in Chapter 8 examine how unequal economic and ecological exchange underlies fish production in Thailand. They find that in the effort to reduce production costs in the fishing industry, slave labor is used in fish harvesting and child and migrant labor is used in processing plants. These highly exploitative operations supply fish to Europe and the United States. They argue that as seafood production shifts with the ongoing growth of aquaculture, fish depletion, and the expansion of fishmeal and fish oil production, the relationships connecting slave labor, slime lines, environmental degradation, and the depletion of marine systems become more deeply embedded in a world-system based on constant capital accumulation that creates socioecological inequalities.

Shellen Wu joins this volume as a historian among sociologists and adds a unique perspective in Chapter 9. Not fully embracing theories of underdevelopment such as Bunker’s, she argues that energy extraction and use in China developed in a very different way than in Europe or in the Amazon Basin of Brazil that Bunker studied in the development of the EUE narrative. Differences resulted from the unique aspects of Chinese geography and geology, as well as China’s response to imperialism from the late nineteenth century. Nor did the so-called *resource curse* strike China, despite considerable Western interest in its coal reserves in the nineteenth century. The case of China discussed by Wu reinforces the importance of history and contingency in any understanding of EUE relations.

The third section of the volume examines responses to patterns of EUE: What has been done? What is to be done? And who should do it? Jackie Smith and Jacqueline Patterson in Chapter 10 examine political activism as it relates to climate change. They argue that real change emerges in movement spaces where people have worked to develop shared perspectives and organization. Specifically, they link their discussion to the earlier theoretical section of the book by highlighting the alternative ontologies and epistemologies being envisioned and forged by the climate justice movement. As they put it, these are not just alternatives but thorough reconceptualizations of identities, values, and social relations. Furthermore, these reconceptualizations rely on indigenous understandings of humanness in relation to/with the earth. By illuminating three examples of transformative projects that are increasingly important—food sovereignty, solidarity economies, and Human Rights Communities—they argue that if the strategies of these projects were widely adopted, then climate change mitigation would be possible.

In the last chapter, David Ciplet and Timmons Roberts also focus on the issue of climate change politics but take a very different route than Smith and Patterson. They analyze the ways in which the Global South is "splintering" over climate change policies. They critique the classic world-systems perspective for dividing the world into a small group of rich countries and a large group of poor, peripheral, and dependent ones. Ciplet and Roberts present rich insights into the series of Conference of Parties (COP) climate change meetings and how they have broken down due to the splintering of previously unified Global South representatives. As they point out, the EUE narrative is increasingly difficult to maintain in the face of climate change politics.

The epilogue of the volume by Harry F. Dahms and R. Scott Frey outlines the major issues discussed in the ten chapters, the implications of these issues, and current gaps in the EUE literature and directions for future theoretical and empirical inquiry.

References

Amin, Samir. 1976. *Unequal Development: An Essay on the Social Formations of Peripheral Capitalism*. New York: Monthly Review Press.

Bunker, Stephen G. 1984. "Modes of Extraction, Unequal Exchange, and the Progressive Underdevelopment of an Extreme Periphery: The Brazilian Amazon, 1600–1980." *American Journal of Sociology* 89(5):1017–1064.

———. 1985. *Underdeveloping the Amazon: Extraction, Unequal Exchange, and the Failure of the Modern State*. Urbana, IL: University of Illinois Press.

———. 2005. "The Poverty of Resource Extraction." Pp. 211–226 in *New Directions in the Sociology of Global Development*, Vol. 11, Research in Rural Sociology and Development, edited by F. H. Buttel and P. McMichael. New York: Elsevier JAI.

———. 2006. *The Snake with Golden Braids: Society, Nature, and Technology in Andean Irrigation*. Lexington, KY: Lexington Books.

Bunker, Stephen G. and Paul S. Ciccantell. 2005. *Globalization and the Race for Resources*. Baltimore, MD: The Johns Hopkins University Press.

Clark, Brett and John Bellamy Foster. 2009. "Ecological Imperialism and the Global Metabolic Rift: Unequal Exchange and the Guano/Nitrates Trade." *International Journal of Comparative Sociology* 50(3–4):311–334.

Emmanuel, Arghiri. 1972. *Unequal Exchange: A Study in the Imperialism of Trade*. New York: Monthly Review Press.

Fischer-Kowalski, Marina. 1998. "Society's Metabolism: The Intellectual History of Material Flows Analysis, Part I: 1860–1970." *Journal of Industrial Ecology* 2:61–78.

Fischer-Kowalski, Marina and Walter Huttler. 1999. "Society's Metabolism: The Intellectual History of Materials Flow Analysis, Part II: 1970–1998." *Journal of Industrial Ecology* 2:107–136.

Foster, John Bellamy. 1999. "Marx's Theory of Metabolic Rift: Classical Foundations for Environmental Sociology." *American Journal of Sociology* 105(2):366–405.

———. 2000. *Marx's Ecology*. New York: Monthly Review Press.

———. 2002. *Ecology Against Capitalism*. New York: Monthly Review Press.

Foster, John Bellamy, Brett Clark, and Richard York. 2010. *The Ecological Rift: Capitalism's War on the Earth*. New York: Monthly Review Press.

Foster, John Bellamy and Hannah Holleman. 2014. "The Theory of Unequal Ecological Exchange: A Marx-Odum Dialectic." *The Journal of Peasant Studies* 41(2):199–233.

Frey, R. Scott. 2015. "Breaking Ships in the World-System: An Analysis of Two Ship Breaking Capitals, Alang-Sosiya, India and Chittagong, Bangladesh." *Journal of World-Systems Research* 21:25–49.

Frey, R. Scott, Paul K. Gellert, and Harry F. Dahms, Guest Editors. 2017. Special issue on Unequal Ecological Exchange in the *Journal of World-Systems Research* 23:226–398.

Georgescu-Roegen, Nicholas. 1971. *The Entropy Law and the Economic Process*. Cambridge, MA: Harvard University Press.

Hornborg, Alf. 1998. "Ecosystems and World Systems: Accumulation as an Ecological Process." *Journal of World-Systems Research* 4(2):169–177.

———. 2009. "Zero-Sum World: Challenges in Conceptualizing Environmental Load Displacement and Ecologically Unequal Exchange in the World-System." *International Journal of Comparative Sociology* 50(3–4):237–262.

———. 2011. *Global Ecology and Unequal Exchange: Fetishism in a Zero-Sum World.* New York: Routledge.

———. 2015. "Why Economics Needs to be Distinguished from Physics, and Why Economists Need to Talk to Physicists: A Response to Foster and Holleman." *The Journal of Peasant Studies* 42:187–192.

Hornborg, Alf and Joan Martinez-Alier, editors. 2016. "Ecologically Unequal Exchange and Ecological Debt." Special issue of the *Journal of Political Ecology* 23:328–491.

Jorgenson, Andrew K. and Brett Clark, editors. 2009a. "Ecologically Unequal Exchange in Comparative Perspective." Special issue of the *International Journal of Comparative Sociology* 50(3–4):1–409.

———. 2009b. "The Economy, Military, and Ecologically Unequal Exchange Relationships in Comparative Perspective: A Panel Study of the Ecological Footprints of Nations, 1975–2000." *Social Problems* 56(4):621–646.

Jorgenson, Andrew K. 2016a. "Environment, Development, and Ecologically Unequal Exchange." *Sustainability* 8(3):227. https://doi.org/10.3390/su8030227.

———. 2016b. "The Sociology of Ecologically Unequal Exchange: Foreign Investment and Environmental Load Displacement: Summary of the Literature and Implications for Sustainability." *Journal of Political Ecology* 23:334–349.

Longo, Stefano B., Rebecca Clausen, and Brett Clark. 2015. *The Tragedy of the Commodity: Oceans, Fisheries, and Aquaculture.* Rutgers, NJ: Rutgers University Press.

Martinez-Alier, Joan, Leah Temper, Daniela Del Bene, and Arnim Scheidel. 2016. "Is There a Global Environmental Justice Movement." *The Journal of Peasant Studies* 43(3):731–755.

Moore, Jason W. 2000. "Environmental Crises and the Metabolic Rift in the World-Historical Perspective." *Organization & Environment* 13(2):123–158.

———. 2011. "Transcending the Metabolic Rift: A Theory of Crises in the Capitalist World-Ecology." *The Journal of Peasant Studies* 38(1):1–46.

———. 2015. *Capitalism in the Web of Life.* New York: Verso.

Oalu, Martin. 2016. "Core Tenets of the Theory of Ecologically Unequal Exchange." *Journal of Political Ecology* 23(1):445–466.

Odum, Howard T. 1971. *Environment, Power, and Society.* New York: Wiley-Interscience.

Patel, Raj and Jason W. Moore. 2017. *History of the World in Seven Cheap Things: A Guide to Capitalism, Nature, and the Future of the Planet*. Berkeley, CA: University of California Press.

Roberts, J. Timmons and Bradley C. Parks. 2007. "Fueling Injustice: Globalization, Ecologically Unequal Exchange and Climate Change." *Globalizations* 4(2):193–210.

Smith, Jackie, Samantha Plummer, and Melanie M. Hughes. 2016. "Transnational Social Movements and Changing Organizational Fields in the Late Twentieth and Early Twenty-First Centuries." *Global Networks* 17(1):3–22.

Wackernagel, Mathis and William E. Rees. 1996. *Our Ecological Footprint: Reducing Human Impact on the Earth*. Philadelphia, PA: New Society Publishers.

Wallerstein, Immanuel. 1974–2011. *The Modern World-System* (Four Volumes). New York: Academic Press.

White, Damian F., Brian J. Gareau, and Alan P. Rudy. 2017. "Ecosocialisms, Past, Present and Future: From the Metabolic Rift to a Reconstructive, Dynamic and Hybrid Ecosocialism." *Capitalism Nature Socialism*. https://doi.org/10.1081/10455752.2017.1296479.

Part I

Theoretical Foundations of Ecologically Unequal Exchange

2

Toward a Theory of Ecologically Unequal Exchange

Stephen G. Bunker

The development or underdevelopment of any region results from the organization, coordination, and use of human and nonhuman energies and from the distribution of resources derived from and transformed in that environment or traded for resources derived from or transformed in other regions. Human uses of any regional environment depend on its ecosystemic characteristics; these are shaped in part by earlier uses and by deliberate human modifications. Social organization, which may enhance or limit access to, and the useful transformation of, natural resources, is both bounded by and further shapes these ecosystems.

Theories of development have focused on economic processes of material transformation, or production, but they have not recognized the absolute dependency of material production on resource extraction. Nor have they accounted for the ways that the extraction, transport, and use

---

From *Underdeveloping the Amazon: Extraction, Unequal Exchange, and the Failure of the Modern State*. Copyright 1985 by the Board of Trustees of the University of Illinois. Used with permission of the University of Illinois Press.

S. G. Bunker (✉)
University of Wisconsin-Madison, Madison, WI, USA
e-mail: rfrey2@utk.edu

R. S. Frey et al. (eds.), *Ecologically Unequal Exchange*,
https://doi.org/10.1007/978-3-319-89740-0_2

of natural resources and the social formations that emerge from these processes affect the subsequent developmental potential of the environments from which resources are extracted. Instead, most theories of development have been attempts to extend models derived from systems of industrial production to nonindustrial systems for which they have only limited relevance.

Recent theoretical literature on national development has compounded the distortions inherent in this bias to production models. Its primary focus has been a fruitless debate about whether the causes of underdevelopment occur in a global system of exchange dominated by industrial nations or within specific regional systems of production. … In fact, a global system of exchange, made up of all importing and exporting regions, determines terms of trade which differentially affect all of these regions, but distinct regional social structures and political arrangements determine how the commodities on which the global system depends are actually extracted or produced. … [We cannot] adequately integrate [competing] perspectives [on national development] unless we recast and incorporate them into ecological and evolutionary models of social change that consider simultaneously the physical dependence of production on extraction and the interaction between regional and global systems. … [R]egardless of the degree to which exchange systems have become global, commodities can emerge only out of locally based extraction and production systems. Models of regional and global systems must be complementary rather than competitive because these systems coevolve. … [D]ifferent regional levels of development result from the interaction between changing demand in the world market for specific commodities and the local reorganization of modes of production and extraction in response. …

The cumulative ecological, demographic, and infrastructural effects of the sequence of modes of production and extraction in any region establish limits and potentials for the productive capacities and the living standards of regional populations. The flow of energy from extractive to productive economies reduces the complexity and power of the first and increases complexity and power in the second. The actions and characteristics of modern states and of their complex and costly bureaucracies accelerate these sequences. Modernization, as ideology, as bureaucratic structure and procedure, and as centralized control through complex regulatory organization, mediates and intensifies the socioeconomic consequences of the interaction between global and regional systems. Modern systems are themselves highly energy-intensive and can only emerge in regions where industrial modes of production derive large amounts of energy and matter from subordinate modes of extraction. The modern state is but one of the forms of social organization which draw on energy flows out of modes of extraction and which extend the dominance of energy-concentrating modes of production, both globally and within nations. I examine these propositions in a case study of the sequence of export economies in the Brazilian Amazon from the time of colonial conquest to the present.

Modes of Extraction and the Creation of Extreme Peripheries

The first essential step toward adequate analysis of the coevolution of regional social formations requires that we free ourselves from notions relevant only to industrial production systems. Concepts derived from the European experience of capital accumulation and technological innovations in industrial production still provide the basic metaphors for the analysis of nonindustrial economies. Economic models of industrial production neglect the extractive origins of the materials which industrial processes transform (Georgescu-Roegen 1975). The internal dynamics of the extractive economies that have provided most of the exports from the least developed regions differ significantly from those of productive economies in their effects on the natural environment, the distribution of human populations, the growth of economic infrastructure (understood here as everything humanly constructed or organized which facilitates social and economic activity), and therefore on the subsequent developmental potential of the affected regions. … [P]roduction models cannot explain the internal dynamics of extractive economies because the exploitation of natural resources uses and destroys values in energy and material which cannot be calculated in terms of labor or capital. When natural resources are extracted from one regional ecosystem to be transformed and consumed in another, the resource-exporting region loses values that occur in its physical environment. These losses eventually decelerate the extractive region's economy, while the resource-consuming communities gain value and their economies accelerate. An adequate model of the interaction between global and regional economies must account for both the differences and the interdependence between the two systems.

The differences between the internal dynamics of modes of extraction and of modes of production create unequal exchange not only in terms of the labor value incorporated into products but also through the direct appropriation of rapidly depleted or nonrenewable natural resources. Extractive appropriation impoverishes the environment on which local populations depend both for their own reproduction and for the extraction of commodities for export. Because this appropriation and its ecological results affect the class structures, the organization of labor, systems of exchange and property, the activities of the state, the distribution of populations, the development of physical infrastructure, and the kinds of information, beliefs, and ideologies which shape social organization and behavior, I introduce the idea, mode of extraction, to suggest the systemic connections between these phenomena. … [B]oth modes of extraction and modes of production can only be understood in terms of their integral interdependence and their impacts on natural ecosystemic processes. Orthodox Marxist notions of the reproduction of modes of production must be reformulated to account for these ecological interdependencies.

While the specific characteristics and dynamics of particular modes of extraction and of particular extractive commodity markets must be analyzed individually, it is possible to outline some general tendencies in extractive export economies. … The extractive process frequently entails an extremely low ratio of both labor and capital to value, so it may initially produce rapid rises in regional incomes. These may be followed by equally rapid collapses when the depletion of easily accessible resources requires additional inputs of labor and capital without corresponding increases in volume. The rapidly rising cost of extraction usually stimulates a search for substitutes or new sources for the original good. Either alternative profoundly disrupts the economy of the exporting region. The ephemeral nature of extractive economies may lead to a series of demographic and infrastructural dislocations.

Productive enterprises typically are located in close proximity to each other. Transportation, communication, and energy transmission costs are thus shared by multiple enterprises. New enterprises can start without assuming the total costs of the infrastructure they require. Populations attracted to these locations provide a labor force which can move easily between enterprises with different rates and directions of growth. While individual enterprises may become obsolete, the infrastructure to which they contribute and the labor which they have employed remain for subsequent enterprise.

Extractive enterprises, on the other hand, must be located in close proximity to the natural resources they exploit. These resources are randomly distributed in relation to productive centers, so proximity to other enterprises occurs only by chance and becomes less likely as the most accessible resources are depleted. Extractive economies, therefore, seldom enjoy the continuities with earlier settlement patterns and infrastructural development which shared productive locations provide. Nor do they usually contribute to the labor and infrastructural requirements of subsequent economies. Instead, whatever changes they bring about in the distribution of population and in the physical environment serve little or no purpose when the specific resources to which they are geared are depleted or are no longer in demand.

Regions whose economic ties to the world system are based almost exclusively on the exchange of extracted commodities (i.e., resources which occur in nature and in whose existence or continued reproduction there is no deliberate human intervention) can be characterized as extreme peripheries because of the low proportions of capital and labor incorporated in the total value of their exports and because of the low level of linkages to other economic activities and social organization in the same region. Even when depletion raises extraction costs, the additional capital and labor are most frequently required for exploration and transport rather than actual extraction. Even then, these costs constitute a relatively small proportion of eventual price, and an even smaller proportion of what their price would be if depletion rates were taken into account (Georgescu-Roegen 1970; Schnaiberg 1980; Schumacher 1973). Examples of such commodities include not only petroleum and minerals but also lumber from natural forests, the oils, meats, hides of wild animals, nuts of undomesticated trees, most fish, and slaves. … [R]aising cattle on pastures formed by burning jungle is also essentially extractive. There is some human intervention in herd management and pasture clearing, but the pasture itself frequently depends on nutrients released from burning vegetation and usually does not last much beyond the rapid depletion of those resources (Hecht 1981).

While processing and industrialization of most extractive commodities create additional value, extreme peripheries such as the Amazon tend to export them raw or unfinished so that the creation and realization of additional values occur in and benefit other economies. Moreover, even the limited contribution of extractive exports to regional economies tends to be unstable; if high demand and expanded scale increase unit costs of extraction by depleting the most accessible resources, entrepreneurs will attempt to domesticate or to synthesize agricultural or industrial substitutes and to transform the extractive economy into a productive one (Brockway 1979). These new economies, once freed from the need to locate near natural resources, will tend to move to areas where land, labor, and infrastructure are more easily accessible.

The crucial difference between production and extraction is that the dynamics of scale in extractive economies function inversely to the dynamics of scale in the productive economies to which world trade connects them. The forces of production develop progressively in industrial systems because the unit cost of commodity production tends to fall as the scale of production increases. In extractive systems, on the contrary, unit costs tend to rise as the scale of extraction increases. Greater amounts of any extractive commodity can be obtained only by exploiting increasingly distant or difficult sources. Though technological innovation may reduce costs of some extractive processes in the short run, unit costs of extraction will continue to rise in the long run. Therefore, when extractive systems respond to increased external demand, they tend to impoverish themselves (1) by depleting non-self-renewing resources or (2) by exploiting self-renewing resources beyond their capacities for regeneration, thereby (3) forcing the unit cost of extracted materials to rise so high that the development of synthetic or cultivated alternatives in other regions becomes cost-effective. These three results are likely to be aggravated by the disruption of the surrounding ecosystem and the consequent reduction of other useful resources whose existence or reproduction depends on biotic chains which include the extracted resource or which are disrupted by the process of extraction.

Successful plantation or industrial production of formerly extractive commodities completes the cycle of peripheral impoverishment by introducing progressive economies of scale in the new location. The competitive advantages of new locations eventually eliminate or seriously reduce the original and increasingly costly extractive economy. In many cases, the economic activities and the settlement patterns which developed around the extractive economy either shrink or become useless. Falling unit costs accelerate production-consumption linkages and infrastructural concentration and accumulation in expanding articulated production systems. The rising unit costs, further dispersion of labor and investment, and intensified ecological disruption which accompany expanding extractive systems eventually decelerate these economies. The intensified energy flow to and through the socially articulated productive systems permits more rapid accumulation there of physical infrastructure, specialized technical and social organizational knowledge in an increased division of labor, and the coordination of research and development of new technologies. These both enhance productive systems' use of nonhuman energy and change the market prices for extracted resources. This capacity to change world markets through technological innovation frees the production systems from short-run dependence on particular extractive commodities as they become depleted and heightens both their dominance over and periodic disruptions of the decelerating modes of extraction.

Production-dominated technological innovations may involve both plant transfers and synthetic substitutions. Brockway (1979) has shown how the development of the botanical and related sciences in the industrial core responded to and promoted the domestication and genetic adaptation of plants extracted in the extreme peripheries to other peripheral regions where center nations controlled both the land and the labor necessary to transform these cultigens into plantation crops. Successful transformation to a plantation system brought these cultigens—rubber, sisal, and cinchona—into a mode of production in which increased scale progressively reduced unit costs to levels at which extractive systems could no longer compete. (These plantation systems frequently aggravated the impoverishment of other extreme peripheries by requisitioning slave or indentured labor.) The incorporation of these extreme peripheries into the world-economy, then, resulted not only in a transfer of value but also in a direct transfer of resources—both natural and human—to less peripheral regions. These plantation systems themselves were finally impoverished by industrial production of synthetic substitutes. Modem searches for oil substitutes—whether nuclear, solar, or agricultural—respond similarly to rising capital and labor costs as the most accessible oil sources are depleted.

Extractive economies tend to develop fewer lateral linkages than productive economies. The well-documented "enclave" nature of extractive economies (see Levin 1960) results from several factors. First, the low proportion of capital and labor to market value concentrates profits in the exchange, rather than in the extractive, sector (Katzman 1976). Second, extractive economies do not respond to the locational advantages that tend to foster the mutual proximity of productive enterprises. Extractive economies necessarily locate at the sources of raw materials, and these sources may be far removed from existing demographic and economic centers. Distance from existing demographic and economic centers increases the costs of labor recruitment, subsistence, shelter, and infrastructural development. In extreme cases, labor is expeditionary, usually involving the temporary migration of males. The additional costs of migration are increased by a near total dependence on imported foodstuffs and other materials, which further reduce the possibility of local economic linkages. This situation in turn enhances control over the labor force, as the provision of subsistence needs is controlled by those who purchase labor. Distance from established communities further heightens the employers' control, as there are few alternative social organizations to provide support for laborers' resistance to exploitation. …

The concentration of capital in removal and transport infrastructure frequently creates especially severe technological dependencies on the industrial countries. Railroads, steamships, docks, drilling rigs, pipelines, and earth-moving machinery require techniques and capital which extractive economies are unlikely to develop. The concentration of investment in export facilities further accentuates the concentrated control over exchange—and profit—which emerges from the absence of alternative economic and demographic linkages. …

Another crucial distinction between extractive and productive modes is that they tend to engender highly distinct, and sometimes contradictory, regimes of land tenure and access to resources. (Distinctions between land ownership and mineral rights in the United States are an example of the different judicial status of production and extraction.) Because extractive location responds to different factors than does productive location, the discovery of valuable resources may well occur on land with no declared ownership, with little or no previous commercial value, and subject to public, rather than private, domain. For all of these reasons, access to resources is of greater import in extractive economies than is actual possession or ownership of land. Rapid increases in the commercial value of natural resources may severely dislocate prior social and economic relations governing possession and use of land, especially when these relations are only tenuously integrated into wider market systems (Bunker 1979). The state, therefore, tends to participate directly in the regulation, authorization, and facilitation of extractive economies.

The tendency for state participation in and facilitation of extractive economies is in many cases enhanced as increasingly difficult access to valuable resources increases extraction costs. The increasing proportion of the Iranian national budget devoted to oil extraction (Fesharaki 1976; Muzegar 1977) or the recent decision of the government of Brazil to invest in the mineral deposits of Carajas (Pinto 1982) may heighten national control of these economies, but they also reduce the state's fiscal capacity to provide social welfare and developmental services in other sectors. …

In contrast to the productive articulated economy, the energy and matter taken from extractive regions do not flow through the extractive economy, do not enhance human productivity or social complexity there, do not engender local production-consumption accelerators, and do not remain embodied in physical infrastructure and complex social organization. The disarticulated extractive export economy can neither generate nor sustain the complex and costly organizational structures of the modem state or the institutions and organizations which the modem state presupposes and on which its functioning depends. The bureaucratic agencies of the modem state can only occur in extractive peripheries as an imposed, exogenous force and are therefore compelled to act without the corresponding civil organization which its own rationality and operating procedures require. … [T]heories of modernization and theories of the authoritarian developmental state provide concepts essential to analyses of these complex bureaucracies, but these theories must be extensively refined to account for the irrationalities and failures of central state intervention in extractive peripheries. …

… Because the social formations of extractive regions seldom develop dense political and economic linkages, and because they lack viable, self-sustaining communities, local inhabitants cannot pressure the state to prohibit repeated disruption through extraction of whatever resources may offer profit to entrepreneurs from other regions. States where the extractive exports predominate in the entire national economy are likely to be more susceptible to pressures and interventions from production-based states, corporations, and cartels than states of nations whose economies are predominantly productive and autocentric.

The particular problem of regions where extractive export economies are predominant is that socioeconomic organization, which at one time responds to international demand for specific extractive commodities, is likely to lose its utility when the extractive source is depleted or when demand shifts away from it. Predominantly extractive economies disrupt human settlement patterns and the natural environment in ways which are adaptive in the short run and maladaptive in the long run. In the absence of self-sustaining and flexible productive systems, there is little or no economic basis for local opposition or resistance to entrepreneurs or to dependent national states that seek to organize the population and environment in such a way as to exploit the potential for quick profit. Thus, extractive economies tend toward eventual stagnation, broken only by new extractive cycles if new demands for new material resources available in the region emerge.

These factors may vary with the characteristics of the national environment, with the type and extent of the national resources extracted, and with the policies of the national state. In the Amazon the tendency for extractive economies at one time to leave the region susceptible to the establishment of subsequent extractive economies (whenever world markets create pressures or opportunities for easy and rapid profits) has led not simply to underdevelopment relative to more rapid increases in productivity in other regions but also to absolute impoverishment and progressive underdevelopment. Where there is little local population to disrupt, however, extractive economies may generate considerable benefits. Valuable minerals or fossil fuels exploited in desert areas with sparse populations may generate revenues which the state can tax or redirect to develop other, productive economies. … Even in these cases, however, the benefits are likely to flow to other areas of the nation where the raw materials are transformed and revenues are directed to more productive enterprise.

Theories of imperialism (Baran 1957; Lenin 1939; Luxemburg 1951), world systems and dependency (see especially Galtung 1971), unequal exchange based on wage or productivity differentials, and modernization have all acknowledged primary material export as a defining characteristic of most forms of underdevelopment, but they have not systematically explored the internal dynamics of extractive systems as a distinct socioeconomic type. Nor have they understood that the complex social organizational, demographic, and infrastructural forms that emerge, as technological change and accumulation accelerate the flow of energy through the articulated productive systems, ultimately depend on processes that progressively decelerate the economy, disrupt the ecosystem, and simplify social organization in extractive regions. None of these theories has accounted sufficiently for the ways in which the extraction and export of natural resources affect the subsequent developmental potential of the environments from which they are extracted.

The extraction of particular commodities from nature has measurable effects on the energy transformation processes in surrounding biotic systems and on the density and distribution of human populations. … [E]xtractive economies tend to "build" the surrounding environment and to distribute human populations in ways which limit, rather than enhance, subsequent forces of production. If this is so, understanding the development and underdevelopment of these environments requires models which systematically take the historical sequence of these effects into account.

Such models are essential to any attempt to reverse or moderate the disruptive effects of extractive economies. It would be pointless to argue against extraction per se because human economic activity and social reproduction cannot occur without extraction. The problem then is to devise ways in which extractive economies can function in a world system of exchange without destroying the social and human environments in which they occur. In order to do so, we must first revalue, theoretically and practically, the natural resources and processes on which economic activity ultimately depends. Theories of value which focus exclusively on labor and capital do not simply err conceptually. Rather, they reflect and legitimate a worldview in which nature is subordinated to mankind and where natural resources are considered flow or income rather than part of a limited global stock or capital (Georgescu-Roegen 1970; Schumacher 1973). Theoretically and practically, nature, values in nature, and the economies which depend on values in nature have been systematically undervalued, while human labor, consumption, and reproduction in articulated societies have been correspondingly overvalued. Revaluing natural energy transformation on a global level would necessarily slow the rate of energy flow from the periphery to the core and, therefore, also slow rates of industrial production and consumption. … [T]his is essential for the long-term reproduction of human society in both extractive and productive modes. Specifying the particular characteristics of and values in extractive economies is an essential first step in any attempt to reverse these economies' disruptive effects, but these characteristics and values must be integrated with more general theories of development and social evolution, both regional and global. …

Energy, Value, and Social Reproduction

The survival and reproduction of society itself must be the ultimate criterion of value, so our concept of value must include anything which affects this process and its outcomes. Labor value, or its imperfect monetary measures, cannot do this. Measures of energy and matter and of their conversion, however, touch everything which is humanly useful. Rather than separating human activity from other ecosystemic processes, these measures allow us to see the interdependencies between human energy use and energy transformation processes which proceed naturally, that is, without human intervention. …

If value is defined only in terms of labor, we have no way to assess the costs which contemporary uses of the environment may impose on subsequent generations and social formations. Energy measures can provide us a calculus of costs and values—past, present, and future—for the multiple effects of human intervention in natural energy transformations. These effects include the disruption of the biotic chains which capture and store energy, the incorporation of energy into immediately consumable goods and services, and the partial conservation of energy and matter in more durable physical infrastructure and social organization.

To use these measures, however, we must reject the anthropocentric and temporally biased notions that value occurs only as a cognitive attribution to certain things or processes. We must also reject the idea that resources do not exist until they are discovered to be useful by humans. Humans may eat fish without knowing or understanding what the fish eat, but this ignorance does not diminish the value of the fish's sources of nutrients to the survival and reproduction of human society. Human activity at one period may destroy or reduce natural energy transformation processes whose usefulness can only be realized with future knowledge or technology; present ignorance does not reduce the cost, or loss of value, to future human generations.

Temporally and culturally bound attributions of value are both socially and epistemologically significant, however, because they affect the allocation and distribution of human labor and the forms of energy extracted from the environment. New technologies and new consumption patterns create new value attributions for the resources they require. The attribution of value to labor enormously influences human decisions about both social organization and uses of natural resources. The differential valuation of labor in different modes of production influences the flow of goods between different economics. Most important, labor is essential to the use value of most naturally occurring resources. All of these, however, constitute only part of the energy transformation processes which sustain human life and society. All of them finally depend on values which occur in nature as the result of energy flows largely independent of human intervention.

Pure extractive economies are the extreme case of human appropriation of these values, but many apparently productive processes include elements of extraction. Different agricultural and pastoral economies, for example, present a gradient of the proportions of human labor and natural values incorporated into the final product, ranging from the minimal modifications of the natural environment in ecologically complex swidden systems to the energy-intensive manipulations and simplifications of bounded ecosystems in large-scale monocropping systems. Forestry exhibits similar gradations.

Human societies depend on complex and variable combinations of natural and labor values. Energy as a measure can be applied to the creation of both kinds of value and allows us to relate them through a common currency. It also allows us to see the usefulness, and thus the value, of human learning and social organization. We can examine the ways that human societies reorganize matter to build their own environments as social inventions which extend the value of portions of the energy which society consumes and dissipates. Finally, it forces us to recognize that there is no possible unidimensional calculus of value because the long-term maintenance of human life depends on energy transformation processes of which we are not yet aware. We cannot measure yet all of the complex energy exchanges in the biotic chains which make up the ecosystems in which we participate. Nor can the value of human organization be directly measured. We know we can use both human and nonhuman energy more effectively because we have remembered past uses of energy and have stored and transmitted this knowledge through social organization, but we could only measure the value of this knowledge and organization by comparing its presence to its absence in the same society. Even without a unidimensional calculus of value, however, we can analyze the very different potentials for social organizational, infrastructural, and economic development in the societies which concentrate energy from outside and the societies which lose energy to them. We can then also explain how the dominance of productive systems accelerates extraction and ecological destruction. …

Most theorists of development and underdevelopment have erred in ignoring the special dynamics and sequences in extractive and other noncapitalist economies. … [T]heir explanations must finally be recast in ecological models of social evolution, but first it is useful to examine them on their own terms. I attempt to correct some of their errors and especially try to show where and how they may fit together. Finally, though, … the decisions and actions of local groups within their own total environment (physical, social, political, and commercial) … [are] essential to the differences between development and underdevelopment of the regions which they inhabit. It is they, after all, who must balance local and global considerations; they who must live with the ecosystems they change; they who must synthesize what are otherwise abstractions of social scientists.

The Internal-External Debate and the Question of Unequal Exchange

… Theories which assign explanatory primacy either to global or to regional systems ignore historical processes, continuities, and dynamics in the other system. … [E]ach system implies a distinct level of analysis and that these levels of analysis must be articulated through a selection of variables and through historical periodizations which provide common referents for both levels of analysis.

Advocates of a global perspective have recently recognized that sociological explanations of underdevelopment must deal simultaneously with two distinct levels of analysis: one appropriate to sequential changes in the socioeconomic structures and processes in particular regionally defined spaces, and the other appropriate to the dynamics of a global system made up of many diverse parts (Frank 1979:2–13; Wallerstein 1981), but their attempts to do this finally founder on their insistence on global levels of analysis. My own strategy is to elaborate a critical synthesis of the externally focused theories of imperialism, dependency, and world system with the internally focused theories of modernization and modes of production, … to determine what parts of existing explanations of development can be usefully incorporated into a more comprehensive ecological model of social and economic change. …

... [In contrast to both internal, modernization theory and external dependency and world-systems theories], the rapid accumulation of capital in the core, which is accelerated by unbalanced energy flows from the periphery, increases the rate of technological and consumption innovation and of consumption capacity in the core. The accelerated production-consumption-accumulation linkages allow the core to determine most global demand. Rapid innovation at the core subjects the periphery to a constantly changing market over which it has little control. If dominant classes in the peripheral areas reorganize modes of production and extraction in response to this externally dominated, frequently shifting market, the populations, social organization, and ecosystems of these areas are subject to repeated disruption. If the local modes of production are not so reorganized, the shifts in demand subject regional economies to falling terms of trade. Wallerstein's metaphorical extrapolations impede attention to these and other regional processes.

... It is precisely because the international market is systemic, that is, the result of the combined production and demand of all of its component modes of production, that exchange in this market and the effects of such exchange on all participating economies must be analyzed as a totality. Actual production systems and their class structure, rates of exploitation, and wage differentials, on the other hand, can finally only be established by separate analysis of specific modes of production. Simply put, we must distinguish between two different levels of exchange. The first involves exchange which occurs between regionally articulated classes and is but one moment of a particular mode of production, or between specific groups in regionally articulated modes of production. The second establishes the global market situations of the various classes which control export and import, and the market-oriented production on which they depend, in all the multiple modes of production which participate in the world system. ...

The use of labor as a standard of value for unequal exchange thus ignores the exchange inequalities inherent in extractive economies, where value in nature is appropriated in one region and labor value incorporated in another. Bettelheim (1972:300–307), for instance, restricts the concept of exploitation to the appropriation of surplus labor value in specific modes of production and thus excludes from consideration the international inequalities involved in the exploitation and export of natural resources. De Janvry (1981:20) extends Bettelheim's restriction in his criticism of dependency and world-system perspectives: "By focusing on the external factors, the underdevelopment school tends to replace the relations of exploitation between social classes with those between geographical areas."

Once we acknowledge, however, that not only the value in labor but also the values in nature can be appropriated, it becomes clear that we cannot counterpose the exploitation between social classes and between geographical areas. Instead, we must consider the effects of the exploitation of labor and the exploitation of entire ecosystems as separate but complementary phenomena which both affect the development of particular regions. We can therefore reject as well Amin's (1977) arguments that unequal exchange occurred only after center wages started to rise above subsistence levels as the result of imperialist strategies which opened world markets and world sources of raw materials for capitalist exploitation. The appropriation of values in nature, from the periphery, in fact initiated unequal exchange between regions, and between ecosystems, long before the rise of wages and the expansion of consumer demand in the core. Examination of the ecological effects of the ivory trade (Palmer and Parsons 1977) and the demographic effects of the slave trade (Wallerstein 1976) on large parts of Africa demonstrates the impact of exploitation between geographic areas as well as between classes on the evolution of unequal exchange.

Additional value is created when extracted materials are transformed by labor. The important point, however, is that this additional value is generally realized in the industrial center, rather than at the periphery. Thus, there are multiple inequalities in international exchange. One, certainly, results from the differential wages of labor. Another, however, is in the transfer of the natural value in the raw resources from the periphery to the center. Another is in the location of the full realization of value and of its accelerated consumption-production linkages in the center, rather than in the peripheral sources of the material commodities. The outward flows of energy and the absence of consumption-production linkages combine with the instability of external demand and with the depletion of site-specific natural resources to prevent the storage of energy in useful physical and social forms in the periphery, and leave it increasingly vulnerable to domination by energy-intensifying social formations at the core. Finally, if the resources do not renew themselves naturally, the inequality of the exchange is intensified by the loss of resources and by the disruption of associated natural energy flows in the periphery itself. …

Finally, there is a tendency in the modes of production approach to attribute "backwardness" (de Janvry 1981) to specific modes of production and to specific class structures. The advantage of a world systemic approach is that "backwardness" is necessarily seen as relational, that is, between connected economies, rather than inherent in a single economy. While it is clearly possible to compare technological and energy consumption levels between economies, societies may also be characterized in terms of their capacity to sustain long-term yields with minimal social inequality. World market participation has severely diminished this capacity in many noncapitalist societies, and the core economies have been able to reduce income inequalities only by intensifying energy flowthrough from the periphery at rates which cannot be sustained over the long run (Adams 1975; de Janvry 1981:17; Georgescu-Roegen 1970; Schnaiberg 1980).

… [W]e [must] recast our economic models to take into account (1) the absolute physical dependence of production on extraction, (2) the locational characteristics and regional inequalities which distinguish productive from extractive systems, (3) the very different ecological, demographic, and social structural evolutionary processes within each type of system, and (4) the long-term consequences of a net flow of matter and energy from extractive to productive economies. The necessary relations between production and extraction, the fact that they typically occur in different regions, and their different ecological results all fundamentally determine both their long-term and short-term potential for social production and reproduction.

Extraction and production may occur together in social formations bounded by a single regional ecosystem. In such cases, the diversity of human needs may distribute extractive activity across such a wide range of species and minerals that biotic chains can reproduce themselves stably. Once the profit-maximizing logic of extraction for trade across regional ecosystems is introduced, however, price differentials between extractive commodities and the differential return to extractive labor stimulate concentrated exploitation of a limited number of resources at rates which disrupt both the regeneration of these resources and the biotic chains of coevolved species and associated geological and hydrological regimes. Once this stage of exploitation has been reached, the industrial modes of production inevitably undermine the resource bases on which they depend. Industrial modes of production have evolved the social organizational and the infrastructural capacity to change their own technologies and thereby to find substitutes for essential resources as they are depleted. This process is necessarily finite, however, as each new technology requires other resources from what is, ultimately, a limited stock.

Analysis of energy flows between regions and of different uses of energy in different regional social formations provides a much fuller explanation of uneven development than any drawn from conventional economic models. If energy and matter necessarily flow from extractive to productive economies, it follows that social and economic processes will be intensified and accelerated in the productive economy and will become more diffuse and eventually decelerate in the extractive economy. The flow of energy and matter to productive societies permits the increased substitution of nonhuman for human energies, allows for increased scale, complexity, and coordination of human activities, stimulates an increasing division of labor, expands the specialized fields of information which this entails, makes possible increasingly complex systems of transport and communication, and engenders the means of technological and administrative innovation by which the crises of resource scarcity are overcome. The mode of extraction, on the other hand, loses energy, and so becomes socially and economically simpler, less diversified, and subject to technologically determined changes in market demand which the modes of production generate. Once we understand this, we can understand as well that, while the actual flow of commodities between regions can be explained in terms of markets and labor costs, the consequent uneven development of different regions of the world can be fully understood only by considering the effects of uneven energy flows on both the physical and social environments of different social formations and on the progressive subordination of simplified, energy-losing societies to increasingly complex, energy-gaining societies....

Our theories of unequal exchange and of uneven development have failed to integrate the internal dynamics of regional social formations with the external dynamic of a world market system because they have not accounted for the necessary relationships between extraction and production or for their consequences on the evolution of different societies. These production-based theories can, however, provide essential components for an ecological model of regionally unequal development, because their basic assumptions closely match the central belief systems of modes of production which currently dominate the world system. Decisions about production, extraction, and exchange are in fact based on anthropocentric value systems which subordinate nature and nonhuman energy to human strategies for enhancing power and control over other humans and for increasing the effective productivity of human labor. These strategies tend to short-term maximization of return to labor and capital with little concern for long-term social reproduction. Conventional theories of development, if properly integrated, can provide us tools to explain the production and exchange decisions and the political and administrative strategies of dominant classes in different kinds of societies. We must go beyond these theories, however, if we wish to understand the consequence of these human decisions for either the short-term development of particular regional social formations or the long-term reproduction of society.

The Problem of Periodization

An adequate theory of development requires that we delineate the "chains of historical causation" (Gutkind and Wallerstein 1976:7) in ways which permit simultaneous reference to both global and regional units of analysis as historically continuous systems. Analysis at the global level has achieved several effective "periodizations" of the world system, but these have all derived from sequential changes in the structure and composition of capital and in the relations of dominant classes to the state in the industrial core (Amin 1974; Baran 1957; Frank 1979; Lenin 1939; Mandel 1975; Preobazhensky 1965). While these periodizations make reference to the impact of these changes on the periphery, they do so by using different peripheral regions to exemplify the dynamics of the different periods (see, e.g., Wallerstein 1976). They thus sacrifice historical continuity at the local level.

I propose that systematic consideration of the commodities exported from a region provides a useful way to periodize both the world and the regional economies and thus to relate the sequence of change in each. … [A] focus on specific commodities permits analysis of the ways that disruption, reduction, or depletion of natural resources may limit the subsequent developmental potential of the environment from which commodities are extracted.

Toward an Ecological Model of Uneven Development

A full account of the intersection between regional and global systems requires separate analysis of each system in terms which recognize the dynamics of each system as an integral unit while simultaneously permitting analysis of their effects on each other. I attempt to achieve this in a historical analysis of the underdevelopment of the Amazon Basin (1) by organizing this history into periods which correspond to the predominance of particular commodities in the Amazon's export trade; (2) by examining the extent to which the combination of political forces and the changes in world-system demand structured the relative composition of exports from the Amazon; (3) by describing how extraction and production of these commodities were organized, either through reorganization of prior modes of production and extraction or through organization of new modes; and (4) by analyzing how the demographic, organizational, infrastructural, and ecological effects of each of these modes of production and extraction established the potential for and the limits on later modes of production and extraction. This articulation of concepts across levels and across time requires precise attention to internal responses to opportunities and pressures generated in external systems. Both the world system and local modes of production and extraction constitute discrete units of analysis whose mutual effects can be seen in the ways that local actors—including those deriving power from organizations that operate beyond the local area—reorganize local modes of production and extraction in order to take advantage of exchange opportunities in the world system.

Treating each local mode of production and extraction as regionally discrete and historically continuous allows consideration of the internal dynamics by which societies may reproduce themselves independently of their participation in a world system and of the variation between societies and over time in participation in, response to, or occasional withdrawal from the world system. This avoids reifying dependency, unequal exchange, or capitalism as causal agents; rather, it permits development and underdevelopment of particular nations or regions to be understood as the ways that particular local classes reorganize modes of production and extraction in response to exchange opportunities and political actions in the world system. …

… Acknowledgment of the specific characteristics of extractive economies and of their differences from productive economies is crucial in this regard, as the geographical and temporal discontinuities of extractive economies are especially likely to lead to discontinuous participation in world systems of capitalist exchange. The human groups which enter and depart from this exchange network are responding to changing market opportunities (see Stavenhagen 1966–67); they maintain themselves through other modes of production and extraction not because capitalism needs them, or because they themselves have become capitalist. Rather, they maintain themselves, as human groups always have, by adapting to their own environments, of which international exchange opportunities form a highly variable part, but which are also structured by the organization of earlier modes of production and extraction. If these adaptations reduce the long-term viability of the physical environment, they also reduce the life chances of the social groups which depend on it.

Georgescu-Roegen (1975) has shown how conventional economic models of production ignore crucial energy transformations which occur between the extraction of material from nature and its use in industry. By focusing only on production, that is, the transformation of these materials by labor and capital, conventional economics ignores the environmental costs of extraction and energy transformation. We will only understand the inequalities inherent in the geographical separation of the different parts of the total processes by which materials in nature are finally transformed for human use and profit when we account for these differential costs to the various regions involved in the world system.

The Amazon: Extractive Exports and Underdevelopment

There are numerous regions of the world whose economic histories would provide relevant cases for a commodity-based model of underdevelopment. The Amazon Basin in Brazil is one of the largest of these regions.

The Amazon Basin has formed an integral part of the world-economy for over 350 years. Soon after Portuguese colonization in the sixteenth century, it was supplying valued spices and animal oils to the European market. From 1860 until 1910, it supplied the bulk of the rubber for the automobiles and other machines which transformed American and European industry. In recent decades, it has supplied increasing proportions of the components for the light metals required by modem transportation technology to reduce the effects of gravity and to reduce fuel consumption. Little of the energy extracted during the Amazon's long history of supplying valued commodities for world trade has been incorporated into enduring and useful social organization and physical infrastructure, however, nor is there much prospect that it will be in the future. On the contrary, the Amazon Basin is one of the poorest areas in the world, and the economic and social systems on which many of its inhabitants depend are seriously threatened by disruption or extinction. This impoverishment continues despite, and in many instances because of, major government development programs. I will examine the utility of my ecological model for explanations of the persistence of such poverty in this huge, resource-rich region.

Effective use and development of natural resources depend on human organization, and the possibilities for effective human organization are bounded by the effects of previous social organization and of previous uses of the environment. The cumulative effects of these sequences on a region's developmental capacities are dramatically illustrated in the case of the Amazon. The decimation of populations during colonial conquest and enslavement, the massive reimportation of human energy to satisfy international industry's needs for rubber in the late nineteenth and early twentieth centuries, and the present expulsion of both peasants and Indians from the lands on which they subsist, in addition to preventing effective and continuous human organization, have been accompanied by increasingly severe depredations of the natural environment. Each depredation, from the killing off of river fauna to the transformation of vast areas of forest into pasture of short economic usefulness and limited capacity for natural regeneration, has severely limited the potential for subsequent human settlement and economic use of the forest.

Sustained economic and social development is impossible when short-term economic and political interests can completely disrupt settlement patterns and the ecological systems on which they depend.... [T]he current exploitation of the Amazon, while lessening the impact of international capital flows and maintaining short-term economic growth for the Brazilian national industrial center, promises to perpetuate the demographic void which previous modes of extraction created in the Amazon's rural areas and thus to restrict its usefulness in the international and national economies of the future.... [T]he energy-intensive and energy absorbing nature of the national state's bureaucracy accelerated the extractive enterprise and the associated disruption of energy flows in the Amazon even when it attempted to reverse these processes. By irrationally extending energy-expensive structures and operating procedures into the energy-poor social formations of Amazonia, the state undermined existing but fragile human communities, devastated the ecosystem in which they subsisted, and severely distorted its own developmental projects. Instead of allowing environmentally balanced strategies for long-term sustained yields, modes of extraction conditioned by politically determined relations of unequal exchange continue to limit the possibilities of social and economic development in the Amazon.

The export economies of the Amazon Basin have been primarily extractive since the colonial period, but the commodities extracted have varied considerably. ... [B]oth the historical sequence and the internal characteristics of these extractive economies account in large part both for the continued poverty of the region and for the fact that it remains, still, an extractive frontier....

Commodity Extraction and Environmental Destruction

The progressive underdevelopment of an extractive periphery organized in response to world market demands is dramatically illustrated by the decimation of indigenous societies and the devastation of key plant and animal resources resulting from colonial exploitation of the Amazon. Where indigenous societies had exploited a wide range of natural energy sources at rates which allowed for their natural regeneration, colonial extraction responded to international demand by exploiting a few highly marketable resources beyond their capacity for natural regeneration, in many cases leading to environmental impoverishments, with widespread ramifications.

The rivers provided a major share of the resource base for dense aboriginal population, but they also provided the avenues for direct European penetration of the most heavily populated areas. As early as the sixteenth century, the Dutch, English, French, Spanish, and Portuguese were struggling to control the Atlantic coast of what is now northern Brazil. Chief among the prizes each sought was control over the sugar producing areas which extended 2000 miles south of the Amazon delta. Secure tenure of these areas depended on control of the river. The earliest permanent penetration started in 1616, when the Portuguese started to build forts to protect national claims to the river basin (Tambs 1974).

Portugal was economically incapable and politically indisposed to finance this military presence in a backwater of a minor colony (Sweet 1974). Both civil and military posts were therefore filled by offering prebendal rights over land and labor *(donatários)*. The spectacular success of sugar plantations on Brazil's Atlantic coast stimulated attempts to implant a similar economy in the Amazon, and the enthusiastic reception in

Europe of native spices *(drogas de sertão)* inspired the organization of extractive expeditions upriver and inland from Belém (Batista 1976; Reis 1975; Sweet 1974). Sugar and spices both required large amounts of manpower. Clearing jungle for monocrops, and planting, harvesting, and processing sugar cane are all extremely labor intensive. The average spice-gathering expedition lasted eight months and required large numbers of Indians as rowers, bearers, hunters, and gatherers (Maclachlan 1973).

Though the Portuguese crown had evidently not intended to repeat its unsuccessful use of Indian labor on sugar plantations (Maclachlan 1973), its practices of granting prebends to its functionaries in the Amazon and drastically limiting the supply of currency there (Sweet 1974) led to extensive enslavement of Indian populations. Colonial production in the Amazon was never profitable enough to support either the purchase of expensive African slaves or the immigration of a European labor force. Slaving expeditions were conducted under a number of pretexts, the most common being that enslaved Indians had been captured in "just wars" or had been ransomed from other Indians who had enslaved them. The threat of slave raids led many Indians to submit to the agricultural-extractive labor regimen of the missions, where they were at least afforded a modicum of protection against such raids (Ross 1978). Even the missions, however, which nominally controlled access to Indian labor, were obliged to make 20 percent of their labor force available for settlers' use, and their ability to enforce restrictions on civil use of Indian labor steadily eroded under political pressure from local government and settlers (Kiemen 1954; Maclachlan 1973; Sweet 1974).

There is evidence that some slavery was practiced among the Amazon tribes prior to the European conquest, but the European demand for "red gold" increased and deepened this practice to the point of severe depopulation. Tribes such as the Tapajo, close to forts and susceptible to constant attack, were held for ransom until they provided slaves of other tribes (Hemming 1978; Nimuendaju 1952; Sweet 1974). There are accounts of the Tapajo turning over their own children when they could not satisfy the European demand through their slave raids on other groups (Nimuendaju 1952). The violence of the slaving raids and the flight of indigenous populations from the fertile river banks, which exposed them to attack, initiated the first, great reduction of native populations (Heriarte 1874).

The demand for Indian slaves and mission labor and the resulting decimation of native populations accelerated as the colonial economy declined. Amazonian sugar could not compete with the sugar plantations on the Atlantic coast in either quality or cost of production. Depletion of the native spices near colonial settlements meant that collecting expeditions had to go farther inland, expanding their need for Indian labor even as European prices fell and local costs rose (Sweet 1974). The slaving expeditions became more and more wasteful of Indian life as the drastic reduction of Indian populations along the rivers increased the time, distance, and expense of slaving expeditions (Reis 1949; Ross 1978). As early as 1693 there were complaints from slavers that it was necessary to go upriver as far as the present boundaries of Peru to find slaves (Hemming 1978). As slaving expeditions had to go farther, they used more Indian rowers and provisions. Due to declining sugar and spice economies, insufficient capital was available to provision the slaving expeditions, so that numerous Indians died from malnutrition on the homebound trip (Sweet 1974). The progressive impoverishment of the colony's natural resource base thus accelerated its decimation of the Indian labor force on which it depended.

The failure of export agriculture heightened the colony's dependence on extraction, and this in turn intensified secular opposition to missionary control of Indian labor. The missions were secularized in 1755; in 1757 the crown stopped encouraging the export of sugar and tobacco, and in 1759 the Jesuits were expelled (Kiemen 1954; Maclachlan 1973). All of this coincided with the establishment and growing power of Companhia Geral do Grão Pará e Maranhão, directed by Governor Mendonca Furtado, brother of the marquis de Pombal. The Companhia's main business was exporting cacao, which grew wild and required the prolonged extractive expeditions which the missions' ability to restrict access to Indian labor would have impeded (Alden 1976; Dias 1970; Herndon 1853; Maclachlan 1973; Ross 1978). The Companhia also exported considerable amounts of lumber, for which it also depended on Indian labor (Alden 1976; Dias 1970).

Infectious diseases brought in by the Europeans may have reduced native population even more than slavery did. The dense riverine populations would have been enormously susceptible to the rapid transmission of new diseases, even ahead of direct contact with Europeans (see Denevan 1976). The combination of crowding, excessive work, and poor nutrition made urban slaves particularly vulnerable to disease. Belem suffered a series of devastating epidemics which ravaged the Indian populations there (Batista 1976; Maclachlan 1973; Sweet 1974). Trade with the missions would also have spread epidemics.

Competition with Dutch and Spanish colonies for territorial and economic control further reduced native populations. Struggles with the Spaniards at the headwaters of the Amazon stimulated conflict between different indigenous groups as well as punitive military expeditions by both colonies (Hemming 1978; Sweet 1974). Dutch manufactured goods were transported up from the Guyana coast and down as far as the middle Amazon in trade between indigenous groups. The Portuguese, worried by this challenge to their monopoly, mounted an extended military campaign against the groups which controlled this trade along the Rio Negro. This campaign culminated in the hunting down, capture, and death of this group's chief, Ajuricaba, and the dispersion of what remained of his group in remote areas of *terra firme* (Sweet 1974).

In addition to direct reduction of native populations, European demand for animal oils eliminated natural resources crucial to the subsistence of dense populations. The manatee, a large aquatic mammal, and the turtle had provided rich supplies of oil and protein for indigenous groups. Smith (1974) has shown that cultural checks and the regionally bounded exchange system maintained a harmonious relationship between turtle and aboriginal populations. Turtles were kept in captivity to balance seasonal fluctuations in other protein sources. They flourished sufficiently in the Amazonian waters to allow their extensive use for meat and oil without population reduction. Pressure from missionaries and early Spanish and Portuguese traders led to massive exploitation of turtle eggs for oil to be sold on local and international markets and of meat for sale as a delicacy much prized by the Europeans.... The manatee was intensely hunted both for local consumption and to supply both oil and meat for ships involved in the West Indies sugar trade.

The rapid reduction of the turtles and the manatees directly deprived the indigenous populations of important sources of oil and meat. It reduced the region's carrying capacity indirectly as well by seriously disrupting critical links in the riverine ecosystem, thus reducing the other riverine resources on which these human populations depended. Turtlings form part of the food chain maintaining the larger fishes, and the manatee's water-surface grazing is crucial in keeping the lakes and channels adjacent to the main river sufficiently free of vegetation to allow the passage of canoes and to permit the entry of light required for the storage of energy in the form of complex organic molecules. Turtles and manatees also stabilize nutrient cycles on which fish depend (Fittkau 1973). As the richest fishing occurs in the quieter waters that are removed from the rivers' main flow, the reduction of the manatee and the turtle greatly diminished the protein resources available to riverine societies.

Finally, the establishment of cattle raising in areas of natural pasture, especially on the *varzea*, the extension of the communities around the various forts, and the later rush for precious minerals pushed the remaining Indians farther away from the more fertile river's edge into the forest where they could only subsist in dispersed and shifting settlements (Bastos 1975; Palmatary 1960).

By the end of the eighteenth century, the twin assaults on native populations and natural resources had created a demographic and economic vacuum, broken only by a few small and impoverished cities. "The existing labor pool had been so overtaxed that no sector, public or private, was able to meet its labor needs" (Maclachlan 1973), even in a depressed economy. The *varzeas* had been almost completely depopulated; much of the technology necessary for their effective exploitation had disappeared with the indigenous societies which had used them (Ross 1978). Europeans had conquered the Amazon, turning those portions of it which had commercial value—Indian labor, turtle and manatee oils and meat, wild spices, and grass—to their own short-term profit in ways which precluded sustained economic exploitation.

The Europeans' rapacity, and their stubborn belief that as members of a master race they should not engage in productive work (Sweet 1974), rapidly exhausted the resources on which their dreams of great wealth were founded. The Amazonian colony sank into unrelieved poverty and stagnation aggravated by political intrigue, frequent epidemics, and the tumult of the years following Brazil's independence from Portugal, when struggles between various ethnic, political, and economic factions eliminated about 20 percent of the total population (Raiol 1970) and devastated the already limited productive capacity which had been developed on the surviving sugar plantations and cattle ranches (Weinstein 1980).

With the exception of sugar and tobacco, little had been produced during two centuries of Portuguese colonization. A great deal, however, had been extracted and sold. The technology used in this extraction was primarily indigenous (Maclachlan 1973), but the core of indigenous productive technology had been lost. Locally dominant classes had established new modes of extraction in response to international exchange opportunities and had used these modes of extraction in ways which decimated local forces of production. The exchange relations in which this class attempted to transform its control over labor and resources into profits were extremely unequal, but conventional calculation of unequal exchange rates based on a labor theory of value or on transfer of surplus to the center is clearly inadequate to analysis of this case. The extraction costs of what was being exchanged included not only human labor but also human life, social organization and technology, and the ecological viability of various interdependent plant and animal systems on which human communities had depended. The effects of this unequal exchange and of the mode of extraction which sustained it directly limited the capacity for local response to and benefit from subsequent exchange opportunities created by industrial development and technological advances in the world system.

Trade strategies designed initially to finance military goals had eliminated effective human occupation and use of most of the Amazon, thus annihilating a previously self-reproducing and sustaining mode of production. When in the mid-nineteenth century Europeans turned to industrial use what many Amazonian Indians had long known—that rubber could be molded into various forms which were both pliant and durable—lack of an adequate labor force retarded the response to a booming new market and was eventually a major factor in the inability of locally dominant classes to organize modes of production adequate to supply and to keep this market. …

Conclusion

The processes which led to and still maintain the underdevelopment of the Amazon can only be understood if we account for the succession of modes of extraction as they emerged from the interaction of regional and global constraints, pressures, and opportunities and as they affected both natural and human environments. None of the prevailing models of development adequately explains these processes. None of the conventional prescriptions for development can be expected to reverse their effects. …

This approach to uneven development allows us to describe more fully than other approaches about the relationships between economic, demographic, social, and ecological processes over time. It allows us to see how, in terms of economic and social growth, uneven development occurs and is maintained. At the same time, however, this approach raises a whole series of questions about the long-term maintenance of industrial modes of production, their effects on extractive regions, and their ultimate vulnerability to resource depletion.

The clearest lesson of class relations in the Amazon is that dominant groups which impoverish the rest of society ultimately impoverish themselves. Only when human communities with balanced exchange relations exist is it possible for social organization to adapt to its total environment in ways which sustain both human community and the ecosystem itself. It is, however, most unlikely that dominant classes will perceive that their long-term interests lie in revaluing human labor and natural resources unless other classes oblige them to understand this.

The point is not that the only solution for a resource-rich region is exit from the world system of exchange. Rather, it is that different regions participate in the world-economy according to exchange opportunities perceived as advantageous by particular classes. … Ultimately, the need is to slow the flow of energy to the world center. As long as natural values in living and fossilized plants which have transformed solar energy into humanly useful forms are transferred predominantly to a small part of the world's total area, the world industrial core will continue to dominate markets in ways which limit the development potential of the rest of the world. … Human groups could, however, use their prescience to enrich, rather than impoverish, the ecosystems in which they participate, both by striving to assure and strengthen natural regeneration and energy transformation processes and by enhancing the effectiveness of their own social organization.

References

Adams, Richard. 1975. *Energy and Structure, A Theory of Social Power*. Austin, TX: University of Texas Press.

Alden, Dauril. 1976. "The Significance of Cacao Production in the Amazon Region during the Late Colonial Period: An Essay in Comparative Economic History." *Proceedings of the American Philosophical Society* 120(April):103–135.

Amin, Samir. 1974. *Accumulation on a World Scale*. New York: Monthly Review Press.

———. 1977. *Imperialism and Unequal Development*. New York: Monthly Review Press.

Baran, Paul. 1957. *The Political Economy of Growth*. New York: Monthly Review Press.

Bastos, A. C. Tavares. 1975/1866. *O Vale do Amazonas*. São Paulo: Companhia Editora Nacional.

Batista, Djalma. 1976. *Complexo da Amazônia*. Rio de Janeiro: Conquista.

Bettelheim, Charles. 1972. "Theoretical Comments." Pp. 271–322 in *Unequal Exchange: A Study of the Imperialism of Trade*. New York: Monthly Review Press.

Brockway, Lucile H. 1979. *Science and Colonial Expansion: The Role of the British Royal Botanic Gardens*. New York: Academic Press.

Bunker, Stephen G. 1979. "Power Structures and Exchange between Government Agencies in the Expansion of the Agricultural Sector." *Studies in Comparative International Development* 14(1):56–76.

de Janvry, Alain. 1981. *The Agrarian Question and Reformism in Latin America*. Baltimore, MD: Johns Hopkins University Press.

Denevan, William M. 1976. "The Aboriginal Population of Amazonia." Pp. 205–234 in *The Native Population of the Americas in 1492*, edited by W. M. Denevan. Madison: University of Wisconsin Press.

Dias, Manuel Nunes. 1970. *A Companhia Geral do Grão Pará e Maranhão*. Belem: Universidade Federado Pará.

Fesharaki, Fereidun. 1976. *Development of the Iranian Oil Industry: International and Domestic Aspects.* New York: Praeger.

Fittkau, Ernst Josef. 1973. "Crocodiles and the Nutrient Metabolism of Amazonian Waters." *Amazoniana* 4(March):103–133.

Frank, Andre Gunder. 1979. *Dependent Accumulation and Underdevelopment.* New York: Monthly Review Press.

Galtung, J. 1971. "A Structural Theory of Imperialism." *Journal of Peace Research* 8(2):81–117.

Georgescu-Roegen, Nicholas. 1970. "The Entropy Law and the Economic Problem." Paper presented in the Distinguished Lectures Series, the University of Alabama, reprinted in *The Ecologist* 2(July):347–381.

———. 1975. "Energy and Economic Myths." *Southern Economic Journal* 41:347–381.

Gutkind, Peter C. W. and Immanuel Wallerstein. 1976. "Editor's Introduction." Pp. 7–29 in *The Political Economy of Contemporary Africa*, edited by Peter C. W. Gutkind and Immanuel Wallerstein. Beverly Hills, CA: Sage.

Hecht, Susanna B. 1981. "Deforestation in the Amazon Basin: Magnitude, Dynamics and Soil Resource Effects." *Studies in Third World Societies* 13:61–110.

Hemming, John. 1978. *Red Gold: The Conquest of the Brazilian Indians.* Cambridge, MA: Harvard University Press.

Heriarte, Maurício de. 1874. *Descrição do Estado do Moranhão, Pará, Corupá, e Rio das Amazonas* (excerpts in Nimuendaju, 1952, and in Palmatary, 1960).

Herndon, William. 1853. *Exploration of the Valley of the Amazon.* Washington, DC: U.S. Navy Department.

Katzman, Marvin. 1976. "Paradoxes of Amazonian Development in a 'Resource Starved' World." *Journal of Developing Areas* 10:445–460.

Kiemen, Mathias C. 1954. *The Indian Policy of Portugal in the Amazon Region, 1614–1693.* Washington, DC: Catholic University Press.

Lenin, V. I. 1939. *Imperialism, the Highest Stage of Capitalism.* New York: International Publishers.

Levin, Jonathan. 1960. *Export Economics.* Cambridge, MA: Harvard University Press.

Luxemburg, Rosa. 1951. *The Accumulation of Capital.* London: Routledge and Kegan Paul.

Maclachlan, Colin. 1973. "The Indian Labor Structure in the Portuguese Amazon, 1700–1800." Pp. 199–230 in *Colonial Roots of Modern Brazil*, edited by D. Alden. Berkeley, CA: University of California Press.

Mandel, Ernest. 1975. *Late Capitalism*. London: New Left Review.
Muzegar, Jahargir A. 1977. *Iran: An Economic Profile*. Chicago, IL: University of Chicago Press.
Nimuendaju, Curt. 1952. "The Tapajo." *Kroeber Anthropological Society Papers* 6:1–12.
Palmatary, Helen Constance. 1960. *The Archaeology of the Lower Tapajos Valley, Brazil*. Philadelphia: Transactions of the American Philosophical Society, New Series, Volume 50, Part 3.
Palmer, Robin and Neil Parsons, editors. 1977. *The Roots of Rural Poverty in Central and South Africa*. Berkeley, CA: University of California Press.
Pinto, Lucio Flavio. 1982. "Amazônia: A Fronteira do Saque." Paper presented at the 31st Annual Latin American Conference, University of Florida, Gainesville.
Preobazhensky, E. 1965. *The New Economics*. Oxford: Claredon.
Raiol, Domingos Antonio. 1970. *Matins Políticos*. Belem: Universidade Federal do Para.
Reis, Arthur C. F. 1949. *Monte Alegre-Aspectos de sua Formaçao Historica*. Belem: Grafica.
Reis, Mauricio Rangel. 1975. *Brasil: 35 Anos de Desenvolvimento*. Brasilia: Ministerio do Interior.
Ross, Eric. 1978. "The Evolution of the Amazon Peasantry." *Journal of Latin American Studies* 10(November):193–218.
Schnaiberg, Allan. 1980. *The Environment: From Surplus to Scarcity*. New York: Oxford University Press.
Schumacher, Ernest F. 1973. *Small Is Beautiful: Economics as if People Mattered*. New York: Harper and Row.
Smith, Nigel J. H. 1974. "Destructive Exploitation of the South American River Turtle." *Association of Pacific Coast Geographers* 36:85–101.
Stavenhagen, Rodolfo. 1966–67. "Seven Erroneous Theses about Latin America." *New York University Thought* 4(Winter):25–37.
Sweet, David. 1974. *A Rich Realm of Nature Destroyed: The Middle Amazon Valley (1640–1750)*. Unpublished Ph.D. Dissertation, Department of Sociology, University of Wisconsin.
Tambs, Lewis A. 1974. "Geopolitics of the Amazon." Pp. 45–90 in *Man in the Amazon*, edited by Charles Wagley. Gainesville, FL: University of Florida Press.
Wallerstein, Immanuel. 1976. "The Three Stages of African Involvement in the World-Economy." Pp. 30–57 in *The Political Economy of Contemporary Africa*, edited by P. C. W. Gutkind and I. Wallerstein. Beverly Hills, CA: Sage.

———. 1981. "On How Accumulation Works." *Contemporary Sociology* 10 (January):41–43.

Weinstein, Barbara. 1980. *Prosperity without Development: The Paraense Elite and the Amazon Rubber Boom (1850–1920)*. Unpublished Ph.D. Dissertation, Department of History, Yale University.

3

Ecologically Unequal Exchange and Raw Materialism: The Material Foundations of the Capitalist World-Economy

Paul S. Ciccantell

The ecologically unequal exchange (EUE) literature analyzes one of the most important pillars of global inequality: the extraction of raw materials from and imposition of environmental damages on peripheral regions, populations, and ecosystems in the capitalist world-economy for the benefit of the wealthy and powerful. This literature fills a glaring hole in world-systems analysis of the 1970s and 1980s: the lack of attention to the role of and consequences for the natural environment in long-term socioeconomic change. The EUE tradition examines commodity chains from their sources through to consumption and waste disposal and often employs sophisticated quantitative analytic techniques to test theoretical models of the causes and consequences of EUE over time and around the world.

This chapter seeks to bring the EUE literature into dialogue with another world-systems theoretical model that integrates global and local natural and social processes over the long term, new historical materialism, or, to put it more bluntly, raw materialism. This theoretical model

P. S. Ciccantell (✉)
Department of Sociology, Western Michigan University, Kalamazoo, MI, USA
e-mail: paul.ciccantell@wmich.edu

R. S. Frey et al. (eds.), *Ecologically Unequal Exchange*,
https://doi.org/10.1007/978-3-319-89740-0_3

focuses attention on the raw materials-based industries and linked transport systems that are used to solve the most fundamental challenge to rapid economic growth: how to acquire growing volumes of raw materials at lower costs and in greater and more secure volumes than other competing economies. One key part of this process of change is how historically many rising economies utilized raw materials access strategies that focused on stealing raw materials peripheries from established hegemons, since the high costs and huge economic political challenges of creating raw materials supply systems had already been paid by the existing hegemon.

The raw materials boom of the 2000s and first half of the 2010s based on China's rapid economic ascent led many firms, politicians, and analysts to see a new "golden age" in which raw materials wealth could serve as the basis for development around the world by tying extractive peripheries to the Chinese market, a world in which the concept of unequal exchange seemed incredibly outdated. The dramatic decline in raw materials prices in 2014–2016 because of the economic slowdown in China is rapidly transforming this golden era into widespread busts for firms and extraction regions. This boom and bust has been particularly dramatic for some mining regions in the coal industry. Because coal has been used for centuries in many locations around the world, coal provides an analytic window into very long-term change. The current historical juncture in China's economic ascent and in coal creates an opportunity for a more robust analysis integrating the insights of ecological unequal exchange and raw materialism to understand the multidimensional causes and consequences of global inequalities.

World-Systems Theory and the Environment

From the intellectual origins of world-systems theory in the work of Wallerstein (1974) until the late 1980s, the environment was at best an afterthought in world-systems analysis. Hopkins and Wallerstein's (1986) initial formulation of commodity chains as an analytical tool noted the role of the natural environment as the starting point of many commodity chains, but this insight received little attention in subsequent work. World-systems analyses of some cases began by noting the natural and environmental characteristics of particular places and industries (e.g.,

Tomich 1990), but these characteristics were largely analyzed as external to the structures and mechanisms of the world-system.

By the late 1980s, a number of scholars in the world-systems and related analytic traditions had begun to incorporate the environment more explicitly into their theoretical models (e.g., Barham, Bunker, and O'Hearn 1990; Bartley and Bergesen 1997; Bergesen and Parisi 1997; Bunker 1985; Ciccantell 1994; Dunaway 1996; Roberts 1992). Building on this growing interest, the 1997 Political Economy of the World-System (PEWS) conference at the University of California-Santa Cruz organized by Walter Goldfrank, David Goodman, and Andrew Szasz (1999) brought together a group of scholars whose work began to place the environment at the center of world-systems analysis.

In the wake of the 1997 PEWS meeting and a special issue of the *Journal of World-Systems Research* that year edited by Bergesen and Parisi (1997), world-systems analyses now often focus a great deal of attention on the natural environment and seek to incorporate the environment into the theoretical framework (for recent efforts, see, e.g., Ciplet, Roberts, and Khan 2015; Kaup 2012; Moore 2015). One of the most influential streams of this environmentally conscious world-systems analysis is EUE; this chapter will now turn to outlining the key tenets of this theoretical and analytical framework.

Ecologically Unequal Exchange Theory

Despite rising interest in the environment in a variety of academic disciplines, public policy debates, and the public recognition of a range of pressing environmental problems in the 1960s and 1970s, environmental sociology remained a largely marginal part of the discipline into the 1990s, even with the creation of institutional bases for the subfield in the American Sociological Society and the Rural Sociological Society (Dunlap, Michelson, and Stalker 2002). Research in the EUE tradition helped move analysis of the environment into a much more central position in the discipline (see, e.g., Jorgenson 2003; Jorgenson and Clark 2009, 2011; Rice 2007). The EUE approach used sophisticated quantitative analytic methods to address world-systems and environmental sociology research questions. This methodological approach provided professional legitimacy to environmentally focused research and to the testing of the theoretical propositions of world-systems theory, allowing publication in leading sociological journals. This EUE methodological approach is part of a broader effort to use sophisticated quantitative analytic techniques to test world-systems theory (see, e.g., Chase-Dunn 1975; Chase-Dunn, Kawano, and Brewer 2000; Kentor 1998, 2001). EUE is grounded in research on global inequalities and their consequences across regions of the world-economy but uses a variety of state-of-the-art sociological methods to address these questions.

In order to identify the key connections between EUE and raw materialism, I emphasize here what I consider to be the essential elements of EUE and the elements most closely linked to the research interests and questions that animate both perspectives. I will expand on these essential elements in subsequent sections.

Five major tenets of EUE are readily apparent in the literature:

1. The foundation of EUE research rests in both world-systems theory and human ecology, focusing analytic attention on examining the appropriation, use, and flows of resources and the resulting waste from this resource use, as well as the myriad environmental impacts of this resource use (Jorgenson and Rice 2012).
2. Unequal material exchange relations, ecological interdependencies, and unequal power across the zones of the world-economy create and reproduce multiple forms of inequality in the world-system (Andersson and Lindroth 2001; Bunker 1985; Hornborg 1998; Jorgenson and Rice 2012; Roberts and Parks 2007).
3. The inequalities of EUE create a seeming contradiction: overconsumption of resources but with relatively less environmental degradation in core countries, while underconsumption of resources in the periphery leaves most residents with poor living and health standards, inadequate incomes, highly polluted shantytowns, and degraded ecosystems (Hornborg 2001; Jorgenson 2003; Jorgenson and Rice 2012).
4. EUE has tremendous negative environmental and social consequences for extractive regions (Bunker 1985; Bunker and Ciccantell 2005; Hornborg 2001; Jorgenson 2003).

5. To demonstrate the operation and impacts of EUE in the world-system, the preferred method is to test theoretical propositions using quantitative analytic techniques (Jorgenson 2004, 2006; Jorgenson and Rice 2005, 2012). Methods used include cross-sectional analysis, Ordinary Least Squares regression, fixed and random effects panel analyses, and network analysis (Jorgenson and Rice 2012).

Over the last two decades, a very large EUE literature has developed; a recent library database search returned a total of almost 200 articles, chapters, and books. Topics examined include climate change (Roberts and Parks 2007; Grimes and Kentor 2003), the role of primary products exports in EUE (Austin 2010, 2012; Jorgenson, Dick, and Austin 2010a), the impact of the military on the environment (Clark, Jorgenson, and Kentor 2010; Hooks and Smith 2005; Jorgenson and Clark 2009), the disposal of various forms of waste in the periphery (Frey 1995, 1998, 2015), deforestation (Burns, Kick, and Davis 2006), the growth of slums in the periphery (Jorgenson et al. 2010a, Jorgenson, Rice, and Brett Clark, 2010b), and coal consumption (Clark, Jorgenson, and Auerbach 2012), among others.

In summary, EUE offers a world-systems theory-based model for understanding the intrinsic interdependence between the capitalist world-economy and the global environment, as well as methodological tools for examining this interdependence. The following section will outline the key features of raw materialist lengthened global commodity chains, and the subsequent section will seek to integrate these world-systems-based approaches.

Raw Materialism and Lengthened Global Commodity Chains

The raw materialist lengthened global commodity chains theoretical and methodological model brings together the global commodity chains model (Bair 2005, 2009; Gereffi and Korzeniewicz 1994; Hopkins and Wallerstein 1986) and new historical materialism (Bunker and Ciccantell 2005, 2007), or, to put it more bluntly, raw materialism. The goal is to move past the dualisms between the global and the local and between nature and society with a theoretical and methodological model that can move between examining, for example, issues as diverse as the impacts of a particular mine on a particular mountain, community, and ecosystem at one point in time and the global geopolitical processes that limit the ability of workers in many locations around the world to organize to achieve economic or other goals during a particular era in the evolution of the capitalist world-economy.

The raw materialist model (termed new historical materialism in the work of Bunker and Ciccantell 2005, 2007) begins from a focus on the material process of economic ascent in the capitalist world-economy. The key problem for rapidly growing economies over the past five centuries has been obtaining raw materials in large and increasing volumes to supply their continued economic development in the context of economic and geopolitical cooperation and conflict with the existing hegemon and other rising economies. Economies of scale offer opportunities to reduce costs and create competitive advantages relative to the existing hegemon and other rising economies, but raw materials depletion and increasing distance create diseconomies of space (increasing costs due to the need to bring raw materials from ever more distant extractive peripheries to the consuming regions) that make finding economic, technological, and sociopolitical fixes via increasing economies of scale difficult to achieve, maintain, and eventually reconstruct on an even larger scale. Successfully resolving this contradiction relies on the creation of generative sectors. Generative sectors create backward and forward linkages; create patterns of relations between firms, sectors, and states; stimulate a range of technical skills and learning and social institutions to fund and promote them; and stimulate creation of a financial system to meet complex and costly capital needs across borders. In short, generative sectors drive economic ascent. Building these generative sectors is a highly contentious and tenuous process that must be maintained in dynamic tension; it is far more common for efforts in rising economies to create and maintain these sectors to fail than to succeed. The United States, Germany, and France all built one of the largest steel industries in the late nineteenth and early twentieth centuries, but only the United States was able to maintain in dynamic tension the material, financial, and political organization that sustained economic ascent to a hegemonic position (Bunker and Ciccantell 2005, 2007).

These processes of economic ascent and economic and geopolitical competition with existing hegemons have driven long-term change in the capitalist world-economy over the past five centuries (Bunker and Ciccantell 2005). The most dramatic and rapid processes of economic ascent restructure national economies and the world-economy in support of national economic ascent. The competitive advantages created by organizational and technological innovations in generative sectors and by subsidies from peripheries lead to global trade dominance. Economic and political competition from the existing hegemon and other ascending economies shapes and constrains long-term success, making economic ascent and challenges to existing hegemons extremely difficult. The most successful cases of ascent restructure and progressively globalize the world-economy, incorporating and reshaping economies, ecosystems, and space. The historical sequence of rapidly ascending economies from Holland to Great Britain to the United States to Japan led to dramatic increases in the scale of production and trade, building generative sectors in iron and steel, petroleum, railroads, ocean shipping, and other raw materials and transport industries that drove their economic ascent and impoverished their raw materials peripheries (Bunker and Ciccantell 2005, 2007).

These processes of long-term structural change, the contradiction between economies of scale and diseconomies of space, and the challenges facing the creation and reproduction of generative sectors raise the bar for future ascendant economies. One central element of the challenges faced by ascendant economies is the recognition that technological and organizational innovations in ascendant economies that resolve this tension simultaneously benefit the rising economy and impoverish its raw materials peripheries, increasing global inequality. Underdevelopment of the periphery (Bunker 1985; Cardoso and Faletto 1969; Frank 1967) is an inherent element of the development of ascendant economies and of existing hegemons.

The lengthened global commodity chains model (Ciccantell and Smith 2009; Sowers, Ciccantell, and Smith 2014) begins analysis of any commodity chain by focusing on raw materials extraction and processing and on the transport and communications technologies that link the multiple nodes of the commodity chain from its raw materials sources through industrial processing to consumption and eventually waste disposal. This approach contrasts sharply with most work in the global commodity chains tradition that focuses on industrial production and consumption and pays little attention to the upstream parts of commodity chains (Ciccantell and Smith 2009). This materially and spatially grounded approach allows analysis of the economic, social, and environmental dimensions of these chains at each node.

Equally important, this approach provides a lens to examine spatially based disarticulations, the marginalization or outright elimination of particular nodes from a global commodity chains, such as via closure of a factory (Bair and Werner 2011), as well as contestations over extraction, processing, transport, consumption, and waste disposal across these chains. This grounded analysis can examine development trajectories and the sociopolitical conflicts over the division of costs and benefits in particular nodes and across these commodity chains. This approach highlights the role of contestation and resistance to the construction and reproduction of a particular commodity chain in particular places, as, for example, labor movements and social movement organizations seek to achieve their goals despite resistance from firms and states that oppose these goals, such as port worker conflicts and the battle over the Keystone XL pipeline and oil from oil sands deposits (Ciccantell and Smith 2009; Sowers et al. 2014).

This model thus emphasizes long-term historical change in the world-system as a whole and in particular places and times, and it allows world-systemic comparative analysis that makes nested and over time comparisons across commodity chains. The grounding in material process also focuses attention on local, regional, and global environmental impacts of these lengthened global commodity chains, a sometime neglected dimension of world-systems analysis. Overall, this raw materialist lengthened global commodity chains approach provides an integrated theoretical and methodological approach to examine the impacts of particular commodity chains both in specific times and places and as the constitutive elements of long-term change in the capitalist world-economy.

Integrating Ecologically Unequal Exchange and Raw Materialist Lengthened Global Commodity Chains

Despite their shared foundations in world-systems analysis and their shared focus on the environmental consequences of human action, EUE and raw materialist lengthened global commodity chains have been largely separate frameworks and methods of analysis. The conference on Ecologically Unequal Exchange at the University of Tennessee, Knoxville, in 2015 gave me an opportunity to interrogate this separation and examine potential areas for dialogue and integration.

One readily apparent point is the numerous shared emphases between the two approaches. In addition to being grounded in world-systems analysis, both approaches seek to move past the nature-society dualism common in the social sciences by emphasizing the inseparability of natural and social processes. Both approaches also analyze how humans use nature, turning some aspects of nature into "resources" useful to humans, disrupting natural processes and changing human processes, solving one challenge by creating others, and so on. Both approaches often examine entire commodity chains and seek to provide an integrated analysis of structure and social action, including contestation and resistance over the creation and maintenance of these commodity chains. Perhaps most fundamentally, both focus on various forms of inequality and their economic, social and environmental consequences.

Given these shared emphases, what are the opportunities for bringing these two approaches into dialogue and potentially integration? Building a large body of explicitly comparative work across a wide variety of commodity chains over time that takes seriously how particular nodes are constitutive of these larger chains will provide a variety of insights that move beyond particular cases, times, and industries to a fuller understanding of both long-term change and how these commodity chains affect particular peoples, locations, and ecosystems. A project is currently underway to develop a methodology to "compare the incomparable" across very different commodity chains (Ciccantell and Smith 2009; Sowers et al. 2014). Given the large body of high quality world-systems commodity chain research (e.g., Bair 2009), one goal of this methodological effort is to provide a means to utilize the cases examined in this literature as data for comparative analysis of a much larger set of global commodity chains. With the ability to utilize data from a large number of diverse cases, many of the quantitative techniques used in EUE research could be employed to analyze this database.

A second area of integration is to utilize the explicitly very long historical time frame of the raw materialist lengthened commodity chains approach to build theoretical models to guide the type of quantitative analyses favored by the EUE approach. Because coal has been a critical component of rising and hegemonic economies for centuries, data on coal production and consumption, economic output, population, and a variety of linked economic, social and environmental indicators are available for a number of countries for decades and, in some cases, centuries. These long-term data again make it possible to utilize some quantitative techniques to examine truly long-term social change linked to the coal industry. We will return to this issue in the following section.

A third area of valuable integration is to emphasize analysis of disarticulation and contestation (Bair and Werner 2011; Ciccantell and Smith 2009; Ciccantell, Sowers, and Smith 2012), particularly over the distribution of costs and benefits of commodity chains for various groups of people and locations, as well as over the long-term sustainability of particular nodes and entire chains. Today, groups struggling against mountaintop removal mining (MTR) in Appalachia and against the expansion of coal mining in Indonesia and other countries are directly contesting continued incorporation of their lands into coal commodity chains that are environmentally unsustainable at local and global scales. This contestation and the resulting potential for disarticulation parallel efforts underway in a variety of other commodity chains and locations (Sowers et al. 2014) and complements EUE research on global climate change and its effects (Clark et al. 2012; Roberts and Parks 2007). Analyzing these struggles and their impacts on commodity chains in conjunction with larger-scale quantitative analysis can form a truly world-systemic analysis across scales.

Constructing and utilizing a truly multidimensional definition and measure of sustainability is a fourth area of potential integration. Global sustainability in the face of climate change, for example, has garnered a great deal of attention, but an integrated approach emphasizes that sustainability is both multidimensional (economic, political, social, and environmental) and multiscalar (global, national, regional, and local). Analyzing multidimensional and multiscalar sustainability across entire commodity chains provides a much more complete and useful tool for understanding the impacts of entire commodity chains and of particular nodes. To show how the integration of the EUE and raw materialist lengthened global commodity chains approaches advances our understanding, I will turn to a discussion of the coal commodity chain and its sustainability in this era driven by China's economic ascent.

The Coal Commodity Chain and Sustainability in the Twenty-First Century

For both EUE and raw materialism, the focus on the relationship between human activity and the environment leads directly to questions of the sustainability of particular forms of human activity, including the use of coal (Bunker and Ciccantell 2005, 2007; Clark et al. 2012). Coal as a case study of sustainability? Who cares about a seemingly anachronistic, very old economy, highly polluting, and dying industry? Coal is in fact an excellent commodity chain through which to study multidimensional and multiscalar sustainability over the very long term. For present purposes, multidimensional sustainability means the need to take into account long-term economic, social, and environmental sustainability of particular commodity chains and their constitutive nodes. Economic sustainability at both the firm and locational levels obviates the need for the ongoing governmental subsidies that have undermined a multitude of economic development efforts around the world. Political sustainability focuses on the debates and contention over public policies that affect multidimensional and multiscalar sustainability of commodity chains, such as the current political debate over climate change in the United States. Social sustainability highlights issues such as employment creation, whether wage levels can sustain workers, families and communities, potential health and safety issues in workplaces, social disruptions of existing communities and boomtowns such as loss of land, drug abuse, and other social problems, and a plethora of other positive and negative potential impacts of incorporation into a particular commodity chain in a particular time and place. Environmental sustainability focuses attention on the typically negative impacts of a commodity chain on local, regional, national, and global environments. Multiscalar sustainability incorporates consideration of these three dimensions at the local, regional, national, and global levels.

Both EUE and raw materialist analyses typically highlight cases of unsustainability, but these theoretical approaches do not presuppose that raw materials extraction is necessarily unsustainable along one or more of these dimensions. For EUE and raw materialism, assessing sustainability and unsustainability are empirical questions to be answered.

One of the most fundamental issues for raw materials-based development in coal or other industries in the increasingly integrated global economy is the economic, political, social, and environmental sustainability of this development. Coal is simultaneously a low-cost, abundant, and essential ingredient for steelmaking and electricity generation, a major contributor to global warming via coal consumption, a cause of ecosystem degradation in areas in which it is extracted and consumed, a topic of global political discussion and debate over the future of the industry, and subject to periods of socioeconomic booms and busts. Given current levels of technological and economic development in the global economy, coal will remain a critical element of steel and electricity production for decades to come, making an understanding of the current and future multidimensional sustainability of this industry a vital issue. In the current era of global economic growth driven largely by the rise of China (Arrighi 2007; Bunker and Ciccantell 2007; Hung 2015), a major producer, importer, and consumer of coal, this issue becomes even more urgent as coal-producing regions such as British Columbia, Australia, and Indonesia increase their exports to China (IEA 2015).

Given this conceptualization of multidimensional and multiscalar sustainability, how can we understand the global coal commodity chain and its constituent nodes? Since the early 1800s, coal has been one of the commodities consumed in greatest total volume, progressively substituting for animals, wood, and wind as a source of heat and power during the nineteenth century. Its contribution to labor productivity and to reduced turnover time and accelerated accumulation of capital has only been possible to the extent that coal was available in great volume at low prices. This combination of high volume and low value meant that coal deposits were the primary determinant of early industrial location, with iron and later steel processing plants and factories that consumed iron and steel located near coal deposits (Harris 1988; Isard 1948). These extractive peripheries subsidized industrialization, consumption, and economic development in core regions, a classic form of unequal exchange and EUE.

However, since the 1950s coal has become one of the most global industries in the world, with 1,383 million tons traded internationally in 2014. How did one of the heaviest, bulkiest, lowest value, and most localized industries in the world get so thoroughly transformed in such a relatively short period of time into one of the largest and most valuable global commodity chains?

For the coal and steel industries, issues of bulk, weight, and transport are the keys to the goal of reducing production costs to make the development of steel and other linked industries globally competitive. The globalization of the coal industry resulted directly from US-led efforts to rebuild Japan after World War II as a geopolitical bulwark in Asia during the Cold War. The US government supported efforts to expand coal production in Australia for export to Japan and helped the Japanese steel firms and the Japanese state create a new model of coastal steel mill and electric power locations that relied on imported metallurgical and steam coal governed by long-term contracts, a model that China has replicated in recent years (Bunker and Ciccantell 2007; Hogan 1999a, 1999b). The US government thus created a new commodity chain that extracted coal and wealth from Australia to subsidize Japan's economic development, leaving coal-extracting regions of Australia with the environmental costs of these subsidies (Bunker and Ciccantell 2007), a classic case of EUE.

The extraction of coal has expanded dramatically in recent decades, as has coal consumption, as the world-economy grew and the coal industry became increasingly globalized. Coal consumption increased spectacularly in China and India in particular in recent years. For the purposes of bringing together raw materialism and EUE, it is important to note that, because of coal's longstanding importance around the world, there are decades and in some cases centuries' worth of data on coal. The International Energy Agency, for example, has published data on coal production, consumption, and trade globally and for most countries covering the last 40–70 years (see Table 3.1).

A few key points are readily apparent in Table 3.1. First, despite growing international concern over the global environmental unsustainability of fossil fuel use, hard coal production has doubled since 2000. Second, much of this increase is due to coal extraction in China. Third, China became the world's largest coal producer in the 1980s and now produces about half of the world's hard coal. The Chinese coal industry dates back centuries (Wu 2015, and Wu this volume) and contributed to China's position as the world's largest and most powerful economy until the 1800s, a position to which China seems to be returning (Frank 1998). Coal production and consumption are helping drive China's economic ascent in the twenty-first century, just as they helped maintain China's economic and geopolitical power in earlier eras.

For longer-term analysis, economic historian B.R. Mitchell (1988, 1998a, 1998b, 1998c) has produced comparable data for many countries around the world on coal production, with annual data for the United States beginning in 1800, for most European countries beginning in the 1810s, for Great Britain beginning in the 1600s, and for China in 1903. Older data on Chinese coal production and use are likely to exist but have not been made available in English. While the beginning dates of

Table 3.1 Hard coal production (millions of metric tons of anthracite, bituminous, and sub-bituminous)

	World	United States	China
1946	1217	481	11
1950	1434	397	42
1960	1990	391	397
1970	2207	550	354
1980	2809	710	620
1990	3566	853	1050
2000	3638	899	1171
2010	6329	926	3140
2016	7268	672	3242

Source: IEA 2001, 2017

these annual tables on the coal industry vary across the regions of the world, it is possible to construct roughly comparable global time-series of one to two centuries in length, a truly long-term basis for analysis.

World coal consumption doubled over the past four decades, as Table 3.2 shows. While consumption has fallen in Organisation for Economic Co-operation and Development (OECD) Europe and has begun to decline in the United States, the tenfold increase in coal consumption in both China and India drove the huge increase in global coal consumption.

Per capita coal consumption (see Table 3.3) has also bifurcated dramatically since the 1980s. In the OECD countries, it has fallen sharply since 1990, while it increased greatly in non-OECD countries since 1990, driven mainly by China and, to a lesser extent, India. As a result, world per capita coal consumption has continued to increase, with 2014 consumption 28% higher than in 1990, despite competition from other fuels (most notably natural gas and renewable sources) for electricity

Table 3.2 Coal consumption (millions of tons of anthracite, bituminous, sub-bituminous, and lignite)

	World	United States	OECD Europe	China	India
1973	3093	505	1056	414	77
1980	3756	650	1157	626	107
1990	4638	815	1155	1049	220
2000	4748	966	817	1337	357
2010	7135	949	749	3221	683
2016	7455	665	671	3610	914

Source: IEA 2017

Table 3.3 World per capita coal consumption (tons of coal equivalent per person)

	OECD countries	Non-OECD countries	World	China	India
1973	1.31	0.30	0.54	0.33	0.08
1980	1.41	0.33	0.57	0.46	0.09
1990	1.43	0.38	0.60	0.65	0.15
2000	1.35	0.36	0.56	0.79	0.2
2010	1.24	0.61	0.73	1.77	0.33
2014	1.13	0.69	0.77	2.07	0.44

Source: IEA 2015

generation and efforts to reduce coal consumption due to concern over global warming.

One major focus of EUE analysis, air pollution based on coal consumption (Clark et al. 2012), is falling in core regions of the United States and Europe, but this ecological and human health cost is exploding in other areas of the world, most notably rapidly industrializing China and India. Growing exports of inexpensive manufactured goods from China to the United States and Europe in recent years rest on a foundation of EUE coal-fired power and steel production and the resulting air and water pollution and human health problems in China.

From the early 1300s onward, England was the dominant force in world coal trade. It was not until the late 1800s that this situation began to change as the United States gained a larger role in this trade and became the leading exporter for much of the post-World War II era. World coal trade has increased even more rapidly than has world coal extraction (see Table 3.4).

China has been following the Japanese model of coastal steel mill development and growing reliance on imported metallurgical and steam coal. China has also been working to steal Japan's raw materials peripheries (Ciccantell 2009; Moyo 2012; Nayar 2004) in coal, iron ore, and other industries. China's hard coal imports remained relatively steady from the 1970s through 2000, but have since exploded, as have India's imports, as Table 3.5 shows.

For Australia and Indonesia in particular and for other coal-exporting countries such as Canada, China's ascent and India's growth and the integration of these extractive peripheries into coal commodity chains linked

Table 3.4 World hard coal trade (millions of metric tons)

	Total	US exports	Australia exports	Canada exports
1960	132	34	1.2	0.77
1970	167	65	18	3.9
1980	263	83	43	15.2
1990	400	95	104	31
2000	594	53	177	31
2010	1076	74	292	33
2016	1333	55	389	30

Source: IEA (1982, 1992, 2001, 2017)

Table 3.5 Coal imports (millions of tons)

	China	India
1960	0.06	0.01
1970	0	0.004
1980	1.99	0.55
1990	2.0	5.1
2000	2.1	24.5
2010	184	121
2016	256	20

Source: IEA (2001, 2017)

to China and India are increasingly making these extractive peripheries look like successful cases of stealing peripheries from earlier ascendants (Ciccantell 2009). From the perspective of economic and social sustainability in these coal mining areas, this process can be seen as beneficial because these coal commodity chains have been growing rapidly and high prices have been partially redistributed to workers and communities in the mining areas. However, from the point of view of global environmental sustainability, this is not the case. In extractive peripheries with less extensive and/or less effective environmental regulation such as Indonesia, this economic opportunity brings with it EUE in the form of forest destruction, water pollution, displacement of existing populations, and dangerous working conditions.

More generally in terms of the sustainability of coal, China's economic slowdown that began in 2014 has led to a decline in demand and prices for coal, raising concerns about the long-term future of coal. European demand has been falling for more than two decades and US demand has begun a sharp decline. However, coal's ready domestic availability in many countries and the existence of a large global trading infrastructure, combined with fears about nuclear power in the post-Fukushima era and the irreplaceability of metallurgical coal for producing primary steel, mean that coal will continue to be consumed in very large quantities for many years to come. The recent climate change accords may hasten the decline of coal in Europe and possibly the United States, but the key locations that shape the future of coal will remain China and India. For some coal-producing countries and coal mining areas within them, coal production is likely to grow, particularly in Australia and Indonesia, providing economic and social sustainability, despite the impacts on local and global environments. For other coal mining areas, especially in Europe and the United States, coal is highly unlikely to be sustainable at any geographic scale or on any of the dimensions of sustainability. In terms of global environmental sustainability, unless energy use patterns change dramatically in China and India, efforts to address climate change are likely to fail. For both EUE and raw materialism, these complex and often contradictory dimensions of sustainability are empirically answerable questions.

Conclusion

How could an integration between EUE and raw materialist lengthened global commodity chains move our understanding forward? One key strategy is developing more powerful methodological approaches that take advantage of the different strengths of the two approaches. Some scholars have built quantitative cross-national analytic models in the world-systems and EUE frameworks (e.g., Jorgenson 2003; Jorgenson and Clark 2009, 2011; Kentor 1998, 2001). Chris Chase-Dunn's work on very long-term change in population, urbanization, and waves of globalization builds on this long tradition of quantitative analysis (Chase-Dunn et al. 2015). There is a lack of long-term data on some topics, but data on a wide variety of industries is available for decades if not centuries (e.g., Mitchell 1988, 1998a, 1998b, 1998c). In the era of Big Data, what else could we do? Computer modeling efforts, for example, Agent-Based Modeling (Gilbert 2008; Miller and Page 2007; Railsback and Grimm 2012), might be another valuable avenue to explore. There might be other methodological options beyond our traditional methods to consider as well.

A second key strategy for using this integrated approach would be to focus on bringing world-systems analysis into more mainstream interdisciplinary socioenvironmental research. The National Science Foundation's Dynamics of Coupled Natural and Human Systems Program (http://www.nsf.gov/funding/pgm_summ.jsp?pims_id=13681), for example, has been open to work using world-systems theory to help understand long-term change. Further, we could work to further move the world-systems perspective into mainstream interdisciplinary environmental research, for example, via the American Sociological Association's Climate Change work (Dunlap and Brulle 2015) that has involved a number of environmental sociologists working from a world-systems-based approach.

A third key strategy is using the analytic power of this integrated approach to inform resistance across boundaries, commodity chains, and nodes. If we are to aid resistance, we must first analyze multidimensional and multiscalar sustainability of particular commodity chains. Then, we have to find ways to work more effectively in the US political context, since resistance currently means not just resistance to climate change and MTR, but also resistance to government inaction in the areas of air pollution and climate change. In terms of the environmental sustainability of coal, human use of coal is a major contributor to climate change; reducing coal consumption appears to be an obvious win for global and national environmental sustainability. Further, the negative environmental and social impacts of coal mining in Appalachia, and especially of MTR coal mining, are also obvious, making it clear that the coal commodity chain is environmentally and socially unsustainable at the regional and local levels in Appalachia. Resistance to MTR in Appalachia is an excellent example of people organizing to contest the power of outside economic and political groups to impose unsustainable social and environmental costs on the people, mountains, and ecosystems of Appalachia. However, resistance in this commodity chain also means resistance to national and international governmental and intergovernmental inaction to address climate change and the current political unsustainability of efforts in the United States to address climate change. If one is inclined to consider this form of resistance irrelevant, it is important to recall that this "War on Coal" political resistance and its influence on West Virginia politics during the 2000 presidential election are the only reason that *Bush v. Gore* in the US Supreme Court mattered in deciding the presidential election.

Fourth, the slowdown and potential bust in China and its impacts on many commodity chains offer a new analytic opportunity for this integrated approach. How will slowing raw materials demand in China affect sustainability across a variety of commodity chains? Can this juncture provide an opportunity to move toward cross-scalar multidimensional sustainability by, for example, allowing mining communities to become sites of less environmentally destructive, employment-generating activities? Could labor organizations and social movement organizations use this transition to build political coalitions to support markedly more sustainable policies in the United States, other countries, and globally?

In sum, integrating EUE and raw materialist lengthened commodity chains to examine multidimensional and multiscalar sustainability in coal and other commodity chains may offer us a way to contribute to efforts to solve some of our most pressing social and environmental challenges in the twenty-first-century world-system.

References

Andersson, Jan and Mattias Lindroth. 2001. "Ecologically Unsustainable Trade." *Ecological Economics* 37:113–122.

Arrighi, Giovanni. 2007. *Adam Smith in Beijing: Lineages of the 21st Century*. New York: Verso.

Austin, Kelly. 2010. "The Hamburger Connection as Ecologically Unequal Exchange: A Cross-National Investigation of Beef Exports and Deforestation in Less-Developed Countries." *Rural Sociology* 75:270–299.

———. 2012. "Coffee Exports as Ecological, Social, and Physical Unequal Exchange: A Cross-National Investigation of the Java Trade." *International Journal of Comparative Sociology* 53(3):155–180.

Bair, Jennifer. 2005. "Global Capitalism and Commodity Chains: Looking Back, Going Forward." *Competition and Change* 9(2):153–180.

———, editor. 2009. *Frontiers of Commodity Chain Research*. Stanford, CA: Stanford University Press.

Bair, Jennifer and Marion Werner. 2011. "The Place of Disarticulations: Global Commodity Production in La Laguna, Mexico." *Environment and Planning A* 43:998–1015.

Barham, Bradford, Stephen G. Bunker, and Denis O'Hearn, editors. 1990. *States, Firms and Raw Materials: The World Economy and Ecology of Aluminum*. Madison, WI: The University of Wisconsin Press.

Bartley, Tim and Albert Bergesen. 1997. "World-System Studies of the Environment." *Journal of World-Systems Research* 3:369–380.

Bergesen, Albert and Laura Parisi. 1997. "Discovering the Environment." *Journal of World-Systems Research* 3:364–368.

Bunker, Stephen G. 1985. *Underdeveloping the Amazon*. Chicago, IL: University of Chicago Press.

Bunker, Stephen G. and Paul S. Ciccantell. 2005. *Globalization and the Race for Resources*. Baltimore, MD: Johns Hopkins University Press.

———. 2007. *An East Asian World Economy: Japan's Ascent, with Implications for China*. Baltimore, MD: Johns Hopkins University Press.

Burns, Thomas, Edward Kick, and Byron Davis. 2006. "A Quantitative Cross-National Study of Deforestation in the Late 20th Century: A Case of Recursive Exploitation." Pp. 37–60 in *Globalization and the Environment*, edited by Andrew Jorgenson and Edward Kick. Leiden, The Netherlands: Brill.

Cardoso, Fernando and Enzo Faletto. 1969. *Dependencies y Desarrollo America Latina*. Mexico City: Siglo Veintiuno.

Chase-Dunn, Christopher. 1975. "The Effects of International Economic Dependence on Development and Inequality: A Cross-National Study." *American Sociological Review* 40(6):720–738.

Chase-Dunn, Christopher, Yukio Kawano, and Benjamin Brewer. 2000. "Trade Globalization since 1795: Waves of Integration in the World-System." *American Sociological Review* 65(1):77–95.

Chase-Dunn, Christopher, Hiroko Inoue, Alexis Alvarez, Rebecca Alvarez, E. N. Anderson, and Teresa Neal. 2015. "Uneven Urban Development: Largest Cities Since the Late Bronze Age." University of California-Riverside Institute for Research on World-Systems Working Paper #98. Available at http://irows.ucr.edu/papers/irows98/irows98.htm.

Ciccantell, Paul S. 1994. *Raw Materials, States and Firms in the Capitalist World Economy: Aluminum and Hydroelectricity in Brazil and Venezuela*. Unpublished Ph.D. Dissertation, University of Wisconsin-Madison.

Ciccantell, Paul S. 2009. "China's Economic Ascent via Stealing Japan's Raw Materials Peripheries." Pp. 109–129 in *China and the Transformation of Global Capitalism*, edited by Ho-Fung Hung. Baltimore, MD: Johns Hopkins University Press.

Ciccantell, Paul S. and David Smith. 2009. "Rethinking Global Commodity Chains: Integrating Extraction, Transport, and Manufacturing." *International Journal of Comparative Sociology* 50(3–4):361–384.

Ciccantell, Paul S., Elizabeth Sowers, and David Smith. 2012. "Are Transport and Raw Materials Still Vulnerable Nodes in Capitalist Commodity Chains?" Paper presented at the 2013 Political Economy of the World-System Section of the American Sociological Association Annual Meeting, Worcester, MA.

Ciplet, David, J. Timmons Roberts, and Mizan Khan. 2015. *Power in a Warming World*. Cambridge, MA: MIT Press.

Clark, Brett, Andrew Jorgenson, and Daniel Auerbach. 2012. "Up in Smoke: The Human Ecology and Political Economy of Coal Consumption." *Organization and Environment* 25(4):452–469.

Clark, Brett, Andrew Jorgenson, and Jeffrey Kentor. 2010. "Militarization and Energy Consumption: A Test of the Treadmill of Destruction Theory in Comparative Perspective." *International Journal of Sociology* 40:23–43.

Dunaway, Wilma. 1996. *The First American Frontier: Transition to Capitalism in Southern Appalachia, 1700–1860*. Chapel Hill, NC: University of North Carolina Press.

Dunlap, Riley and Robert Brulle, editors. 2015. *Climate Change and Society*. New York: Oxford University Press.

Dunlap, Riley, William Michelson, and Glenn Stalker. 2002. "Environmental Sociology: An Introduction." Chap. 1 in *Handbook of Environmental Sociology*, edited by Riley Dunlap and William Michelson. Westport, CT: Greenwood.

Frank, Andre Gunder. 1967. *Capitalism and Underdevelopment in Latin America*. New York: Monthly Review Press.

———. 1998. *ReOrient: Global Economy in the Asian Age*. Berkeley, CA: University of California Press.

Frey, R. Scott. 1995. "The International Traffic in Pesticides." *Technological Forecasting and Social Change* 50:151–169.

———. 1998. "The Hazardous Waste Stream in the World-System." Pp. 84–103 in *Space and Transport in the World-System*, edited by Paul S. Ciccantell and Stephen G. Bunker. Westport, CT: Greenwood Press.

———. 2015. "Breaking Ships in the World-System." *Journal of World-Systems Research* 21(1):25–49.

Gereffi, Gary and Miguel Korzeniewicz, editors. 1994. *Commodity Chains and Global Capitalism*. Westport, CT: Praeger.

Gilbert, Nigel. 2008. *Agent-Based Models*. Thousand Oaks, CA: Sage.

Goldfrank, Walter, David Goodman, and Andrew Szasz, editors. 1999. *Ecology and the World-System*. Westport, CT: Greenwood Press.

Grimes, Peter and Jeffrey Kentor. 2003. "Exporting the Greenhouse: Foreign Capital Penetration and CO_2 Emissions 1980–96." *Journal of World-Systems Research* 9:261–275.

Harris, J. R. 1988. *The British Iron Industry 1700–1850*. London: Palgrave.

Hogan, William. 1999a. *The Steel Industry of China: Its Present Status and Future Potential*. Lanham, MD: Lexington Books.

———. 1999b. "The Changing Shape of the Chinese Industry." *New Steel* 15(11):28–29.

Hooks, Gregory and Chad Smith. 2005. "Treadmills of Production and Destruction: Threats to the Environment Posed by Militarism." *Organization and Environment* 18(1):19–37.

Hopkins, Terence K. and Immanuel Wallerstein. 1986. "Commodity Chains in the World-Economy Prior to 1800." *Review: A Journal of the Fernand Braudel Center* 10:157–170.

Hornborg, Alf. 1998. "Ecosystems and World Systems: Accumulation as an Ecological Process." *Journal of World-Systems Research* 4(2):169–177.

———. 2001. *The Power of the Machine: Global Inequalities of Economy, Technology, and Environment.* New York: AltaMira Press.

Hung, Ho-fung. 2015. *The China Boom: Why China Will Not Rule the World.* New York: Columbia University Press.

International Energy Agency (IEA). Various Years. *Coal Information.* Paris: Organization for Economic Cooperation and Development.

Isard, Walter. 1948. "Some Locational Factors in the Iron and Steel Industry Since the Early Nineteenth Century." *Journal of Political Economy* 63(3):203–217.

Jorgenson, Andrew. 2003. "Consumption and Environmental Degradation: A Theoretical Proposition and Cross-National Study of the Ecological Footprint." *Social Problems* 50(3):374–394.

———. 2004. *Export Partner Dependence and Environmental Degradation, 1965–2000.* Unpublished Ph.D. Dissertation, Department of Sociology, University of California-Riverside.

———. 2006. "Unequal Ecological Exchange and Environmental Degradation: A Theoretical Proposition and Cross-National Study of Deforestation, 1990–2000." *Rural Sociology* 71:685–712.

Jorgenson, Andrew and Brett Clark. 2009. "The Economy, Military, and Ecologically Unequal Relationships in Comparative Perspective: A Panel Study of the Ecological Footprint of Nations, 1975–2000." *Social Problems* 56:621–646.

———. 2011. "Societies Consuming Nature: A Panel Study of the Ecological Footprints of Nations, 1960–2003." *Social Science Research* 40:226–244.

Jorgenson, Andrew, Christopher Dick, and Kelly Austin. 2010a. "The Vertical Flow of Primary Sector Exports and Deforestation in Less-Developed Countries: A Test of Ecologically Unequal Exchange Theory." *Society and Natural Resources* 23:888–897.

Jorgenson, Andrew and James Rice. 2005. "Structural Dynamics of International Trade and Material Consumption: A Cross-National Study of the Ecological Footprint of Less-Developed Countries." *Journal of World-Systems Research* 11:57–77.

———. 2012. "The Sociology of Ecologically Unequal Exchange in Comparative Perspective." Pp. 431–439 in *Routledge Handbook of World-Systems Analysis*, edited by Salvatore Babones and Christopher Chase-Dunn. New York: Routledge.

Jorgenson, Andrew, James Rice, and Brett Clark. 2010b. "Cities, Slums, and Energy Consumption in Less-Developed Countries, 1990–2005." *Organization and Environment* 23:189–204.

Kaup, Brent. 2012. *Market Justice: Political Economic Struggle in Bolivia*. Cambridge: Cambridge University Press.

Kentor, Jeffrey. 1998. "The Long-Term Effects of Foreign Investment Dependence on Economic Growth." *American Journal of Sociology* 103(4):1024–1046.

———. 2001. "The Long-Term Effects of Globalization on Income Inequality, Population Growth, and Economic Development." *Social Problems* 48(4):435–455.

Miller, John H. and Scott E. Page. 2007. *Complex Adaptive Systems: An Introduction to Computational Models of Social Life*. Princeton, NJ: Princeton University Press.

Mitchell, B. R. 1988. *British Historical Statistics*. Cambridge: Cambridge University Press.

———. 1998a. *International Historical Statistics: Africa, Asia and Oceania 1750–1993*. New York: Macmillan Reference.

———. 1998b. *International Historical Statistics: Europe 1750–1993*. New York: Macmillan Reference.

———. 1998c. *International Historical Statistics: The Americas 1750–1993*. New York: Macmillan Reference.

Moore, Jason. 2015. *Capitalism in the Web of Life*. London: Verso.

Moyo, Dambisa. 2012. *Winner Take All: China's Race for Resources and What It Means for the World*. New York: Basic Books.

Nayar, Baldev Raj. 2004. "The Geopolitics of China's Economic Miracle." *China Report* 40(1):19–47.

Railsback, Steven and Volker Grimm. 2012. *Agent-Based and Individual-Based Modeling: A Practical Introduction*. Princeton, NJ: Princeton University Press.

Rice, James. 2007. "Ecological Unequal Exchange: International Trade and Uneven Utilization of Environmental Space in the World System." *Social Forces* 85:1369–1392.

Roberts, J. Timmons. 1992. *Forging Development, Fragmenting Labor: Subcontracting and Local Response in an Amazon Boomtown*. Unpublished Ph.D. Dissertation, Department of Sociology, Johns Hopkins University.

Roberts, J. Timmons and Bradley Parks. 2007. *A Climate of Injustice: Global Inequality, North-South Politics, and Climate Policy*. Cambridge, MA: MIT Press.

Sowers, Elizabeth, Paul S. Ciccantell, and David A. Smith. 2014. "Comparing Critical Capitalist Commodity Chains in the Early Twenty-first Century: Opportunities for and Constraints on Labor and Political Movements." *Journal of World-Systems Research* 20(1):112–139.

Tomich, Dale. 1990. *Slavery in the Circuit of Sugar: Martinique in the World Economy, 1830–1848*. Baltimore, MD: Johns Hopkins University Press.

Wallerstein, Immanuel. 1974. *The Modern World-System I: Capitalist Agriculture and the Origins of the European World-Economy in the Sixteenth Century*. New York: Academic Press.

Wu, Shellen Xiao. 2015. *Empires of Coal: Fueling China's Entry into the Modern World Order, 1860–1920*. Stanford, CA: Stanford University Press.

4

The Role of the Semi-Periphery in Ecologically Unequal Exchange: A Case Study of Land Investments in Cambodia

Mariko Frame

Large-scale land investments have increased dramatically in the past decade, attracting attention from scholars and alarming activists and civil society organizations. In many cases, this trend has been accompanied by widespread evictions and dispossessions, civil unrest, and environmental degradation (Borras et al. 2012; Hall 2011; Klare 2012; Oakland Institute 2009, 2011; Visser, Mamonova, and Spoor 2012). Cambodia is one of the most extreme examples, where land investments have displaced hundreds of thousands of Cambodian citizens and deforested vast tracts of Southeast Asia's most biodiverse forests. The history of capitalism has been built upon the dispossession of land (Magdoff 2013), but in the era of globalized capitalism and ecological crises 'land grabs' in Cambodia and elsewhere are partially driven by investors from industrializing countries in the Global South.

The broader theoretical question that emerges is how to conceptualize the role of such emerging economies in the global ecology/world-system. One of the foremost theories that attempts to explain how global

M. Frame (✉)
Global Studies, Politics and Justice Program, Westminster College,
Salt Lake City, UT, USA

R. S. Frey et al. (eds.), *Ecologically Unequal Exchange*,
https://doi.org/10.1007/978-3-319-89740-0_4

environmental burdens are distributed across the international division of labor is ecologically unequal exchange (EUE). EUE derives from dependency and world-system theories, whereby the global economy is seen as a world-system in which economic tasks are geographically distributed across a core and periphery. Hornborg (2011) argues that the industrial metabolism of core centers necessitates the appropriation of peripheral hinterlands' resources and the usurpation of peripheral sink-capacity. EUE therefore posits an asymmetrical transfer of resources between core and peripheral economies (Hornborg 1998, 2001, 2011). Such an exchange leads to a systematic deterioration in the ecological situation of peripheral economies (with the associated social and economic losses) to the benefit of core economies (with the associated gains in high levels of resource consumption, capital accumulation, and technological advance) (Bunker 1985). The perpetuation of the flow of these uneven ecological burdens is predicated on the capacity of core industrial nations to engage in trade, or other forms of exchange, with peripheral nations that is fundamentally 'ecologically unequal.'

The issue is that such emerging economies as the ones in this small study are neither properly core nor peripheral. Instead, the central argument of this chapter is that such emerging economies occupy a distinct, intermediate semi-peripheral status, and as such face a number of complex contradictory dynamics and distinct pressures that drive ecological degradation both within and outside of their national borders. Specifically, such countries face an intense upward competition in a hierarchical capitalist global economy that pressures them to industrialize rapidly, resulting in environmental exhaustion of their own domestic natures. In addition, such countries continue to engage in peripheral activities that are centered on the extraction and export of primary commodities, and such activities also drive domestic environmental degradation. In response to both of these dynamics, as well as their increasing levels of consumption resulting from economic growth, these emerging economies are increasingly seeking to externally secure land and resources. Furthermore, this chapter argues that this trend is occurring under a form of regional ecological imperialism.

The other central point this chapter makes is that even as some semi-peripheral countries are engaging in forms of ecological imperialism with peripheral countries, they remain subordinate to core economies in a number of ways. To begin with, such countries do not occupy the same position as core countries in terms of the global hierarchy of wealth, even if they are industrializing. Second, in many cases semi-peripheral economies operate as suppliers and subcontractors to corporations in the Global North, and many of their exports end up consumed in the Global North. Third, international financial institutions of the Global North, particularly the World Bank and International Monetary Fund, have been instrumental in promoting the industrialization and neoliberalization of semi-peripheral natures. They have also been instrumental in opening up peripheral economies and allowing the appropriation of their natures by both core and semi-peripheral countries. For all these reasons, this chapter finds that in terms of EUE such countries do not fit into the binary category of core versus periphery which has characterized EUE theory thus far. Rather, the emerging economies under study appear to both exploit and be exploited.

The first section of this chapter offers some theoretical explanations as to why the role of the semi-periphery is a necessary category in the theoretical formulation of EUE. The second section clarifies the connection and distinction between EUE and ecological imperialism. The rest of this chapter looks more closely at the case study of East Asian and Southeast Asian investment in Cambodia's land sector as a form of ecological imperialism.

The Importance of Semi-Periphery Status in the Global Ecology/Economy

To a large extent, the issue of EUE is an empirical issue; either the world-system is characterized by asymmetrical flows of resources from peripheral regions to core regions, or it is not. While many empirical studies corroborate EUE between developed and less-developed countries, some studies have found that due to tremendous economic growth, rising consumption standards, and increasing demands for natural resources, certain 'emerging' economies have become, or are becoming, net importers of resources. In world-systems theory, such countries could be considered part of the 'semi-periphery.' World-systems theory has long analyzed semi-peripheral trends of economic growth and industrialization (Arrighi 1990; Arrighi and Drangel 1986; Arrighi, Silver, and Brewer 2003; Chase-Dunn 2005; Ruvalcaba 2013; Wallerstein 1974). However, while recent empirical work on EUE has included middle-income countries and emerging economies (Singh and Eisenmenger 2010; UNEP 2013), the semi-periphery remains under-theorized in the EUE literature. That is, theoretical work on EUE is still largely couched in the binary terms of core and periphery, developed versus less developed, industrialized versus periphery, and so on. For example, EUE has been defined by Rice (2009:221, emphasis added) as

> the environmentally damaging withdrawal of energy and other natural resource assets from the periphery and the addition or externalization of environmentally damaging production and disposal activities within the *periphery* of the world system. It constitutes both the obtainment of natural capital or the stocks of natural resources that yield important goods and services and the usurpation of sink-capacity or waste assimilation properties of ecological systems in a manner enlarging the domestic carrying capacity of *industrialized* countries to the detriment of *peripheral* societies.

In a more recent piece, Jorgenson (2016) adopts this definition, again using the terminology of 'developing countries,' 'less-developed countries,' and 'industrialized countries.' Hornborg (1998, 2001, 2011) also relies largely on the categorizations of 'core' or 'industrial centers' and 'peripheral hinterlands.'

Such broad generalizations do not necessarily exclude the possibility of EUE between more industrialized countries within the Global South and less industrialized countries.[1] However, for a number of reasons this chapter discusses, this exchange is qualitatively different because semi-peripheral countries do not occupy the same position within the world-system as core countries do, in terms of either wealth or power. In world-systems theory, the semi-periphery plays a distinct role. Chase-Dunn and Hall (1997:78) define the semi-periphery as:

1. Regions that mix both core and peripheral forms of organization.
2. Regions spatially located between core and peripheral regions.

3. Regions spatially located between two or more competing core regions.
4. Regions in which mediating activities linking core and peripheral areas take place.
5. Regions in which institutional features are intermediate in form between those forms found in adjacent core and peripheral areas.

Core countries are characterized by highly developed forces of production, including high-paid labor, capital-intensive production, and technological advancement. Accordingly, they appropriate a disproportionate share of the benefits of the international division of labor. In contrast, peripheral countries are characterized by low-level forces of production, including low-paid labor and labor-intensive production, particularly the export of raw materials. Semi-peripheral countries tend to contain a mixture of both core and peripheral economic activities, and while able to appropriate a relatively greater share of the global surplus than the periphery, nonetheless appropriate less than core countries. According to Chase-Dunn and Hall (1997), they are both dominated by core countries and dominate peripheral zones.

The world-systems conception of the world-economy as a hierarchy remains salient for our understanding of global capitalism and the global ecology, even if the categories of core, semi-periphery, and periphery are rough approximations at best. For one, there has been a rush of mainstream literature in recent years underscoring the rising economic competitive threat of China and other emerging economies, and a corresponding set of literature condemning their rush of resource and land grabbing.[2] While the socio-ecological effects of semi-peripheral industrialization trends are known and discussed, the mainstream literature tends to ignore the continued underlying hierarchy within the world-system and its implications.

As the works of Arrighi et al. (2003) have emphasized, what matters is not so much the particular mix of economic activities in defining what is core, semi-peripheral, or peripheral, but *the capability of a state to appropriate the benefits of the world division of labor*. This is determined, they argue, primarily by a state's position in the global hierarchy of wealth. The further up in the hierarchy of wealth a state is, the better positioned its rulers and subjects are in the struggle for benefits. While semi-peripheral states may be industrializing, major profit-oriented innovations, and the profits associated with them, tend to cluster in 'zones of prosperity' within core countries. In contrast, by the time new products and techniques are adopted by the poorer countries they tend to be subject to intense competition and no longer bring the high returns they did in wealthier countries. Arrighi (1990) found for the 1938–1983 period a stable hierarchy of wealth in the global economy. He identified, based on the world distribution of gross national product per capita, three distinct clusters of countries (high, middle, and low) and found that long-term upward/downward mobility of countries from one cluster to another was exceedingly rare. In other words, despite changes in levels of industrialization throughout the semi-periphery and periphery, the capacity of non-core states to capture the global surplus did not change significantly.

Alternatively, one could conceptualize this issue from the perspective of global commodity chains. The capitalist world-economy is connected through millions of chains of production and exchange, from extraction through final consumption (Ciccantell and Smith 2009). Within each node of each commodity chain, a certain amount of surplus is generated. However, the surplus generated is unevenly distributed among the states in the world-system, so that even developing countries with large economies such as China are not able to secure as large a surplus as the developed countries, reflecting their semi-peripheral, or even peripheral, status (Li 2003). Analyzing multiple network analysis of various types of international commodity trade between 1965 and 2000, Smith and Mahutga (2009) found that while it is true that some newly industrializing countries have successfully engaged in industrial upgrading in which there is a shift from commodities such as textiles, apparel, and footwear to higher value-added items, many peripheral countries remain primarily export platforms for low-technology, labor-intensive goods made by low-wage, unskilled workers. They argue that their findings suggest that the structure of the world-economy still conforms to the overall core-semi-periphery-periphery layering.

In EUE theory, the peripheral hinterland is seen as the source and sink for the metabolic needs of the core industrial centers, bolstered by unequal exchange of resources under the nominally fair monetary exchanges of international trade. In such a formulation, two central issues appear immediately significant to the study of EUE. First, we need to determine where (geographically) resources are being extracted and where wastes and pollution are being discarded. From this we can determine which countries primarily shoulder the environmental burdens of global industrial capitalism. That is, we can determine the 'ecologically unequal' aspect. The second issue is equally important: where profits accumulate. Where profits accumulate determine both the capacity of certain countries to consume a relatively higher share of world resources, as well as their continued capacity to appropriate such resources for further investment, production, and accumulation of capital. Both of these issues determine the relationships of power and exploitation between countries in the hierarchy of the world-system.

For the reasons stated above, any simple core–periphery dichotomy, both in terms of geography of environmental burdens and the geographical location of accumulated capital, is insufficient. As world-systems theorists argue, the world-system is at least three-tiered, not two-tiered, in terms of the hierarchy of wealth. Further, the growth and industrialization of emerging economies are continuously re-arranging many aspects of the global economy/ecology that concretely affect the direction of biophysical flows. Many semi-peripheral states are now net importers of resources and engaging in environmentally destructive trade and foreign investment relations with the periphery. Yet importantly, reflecting their intermediate status within the world-economy and their mixture of both peripheral and core economic activities, semi-peripheral countries both exploit peripheral natures and continue to be exploited by core regions. In a world-system characterized by the capitalist mode of production, such dynamics are driven by the expansionary, ceaseless pursuit of capital accumulation and through an international division of labor that remains hierarchical, but not necessarily static.

Ecologically Unequal Exchange and Ecological Imperialism

Theorists of imperialism have long emphasized the necessity of geographical expansion and appropriation of peripheral resources as a requisite for the ceaseless pursuit of capital accumulation, though different theorists emphasize different dynamics. Magdoff's (2003) work, for example, emphasizes technological changes and competition between capitalist powers. Moore (2001, 2003, 2011, 2012) asserts that capitalism has sustained itself on the basis of cheap inputs and through mobilizing a succession of great leaps forward in the extraction of ecological surplus. Such dynamics necessarily push expansionary tendencies on a global scale, as they typically result in the rapid exhaustion of nature (and human nature) which can act as fetters on the accumulation of capital. Re-formulating Marx's original concept of metabolic rift, Foster, Clark, and York (2011) argue that capitalist social relations of production, defined as the separation of the workers from their means of production, are both cause and condition of a historically distinct and environmentally destructive capitalist geography (see Gellert, Chap. 5, this volume). Clark and Foster (2009), actually utilizing the term 'ecological imperialism,' link the issue of imperialism with EUE in the specific case study of the international guano trade of the nineteenth century.

Drawing from the works of the above theorists, I have identified some essential attributes of ecological imperialism (Frame 2014, 2015). Centrally, ecological imperialism is rooted in the endless drive for capital accumulation and occurs under specifically capitalist relations of production. There are numerous specific dynamics driving ecological imperialism—ranging from competition between capitalist powers, dynamics of the metabolic rift at a global scale, the role of finance and monopoly capital, among others all worthwhile of in-depth study—but at their root is the drive for capital accumulation. Second, ecological imperialism hinges upon dynamics of unequal power (economic, political, military, ideological, and so on) and dependency within a hierarchical international division of labor in the world-system, as a historical result of colonialism and uneven development. Inequality and certain class structures are also crucial in upholding the political pillars of imperialism. Third, ecological imperialism results in negative socio-ecological impacts for peripheral countries. Overall, ecological imperialism allows for the displacement of environmental burdens outside of core/semi-peripheral national borders.

Beyond these attributes found in the existing literature, the form ecological imperialism takes is an outcome of the dialectical unfolding of movements and countermovements that arise in response to social resistance, within the periphery, but also within core and semi-peripheral countries. Imperialism is dynamic and responsive, it does not just hinge on economic factors but also technological, social, political, and ecological factors. Ecological imperialism necessitates an amenable political-economic context within the imperialized country. And, historically it has been met with resistance at certain times. Hence, the price of nature in the periphery depends not only on market supply and demand and ground rent but also on class, anti-imperialist, and environmental struggles. Currently, the neoliberal reforms driven by international financial institutions (IFIs) that began in the 1980s have rendered peripheral resources once again cheap and easily accessible as they were under colonialism and neocolonialism. Such neoliberalization of nature policies, in which nature is enclosed, privatized, valuated, and marketized, is the current form of ecological imperialism in the peripheral regions (Frame 2016).

Overall, then, we can broadly identify ecological imperialism as the subjugation of the economic, political, and/or social institutions of a (generally peripheral) country for the biophysical, metabolic needs of the (generally the core or semi-periphery), inextricable from the purpose of making such resources accessible and amenable (in the right quantities and for the right price) to the needs of (foreign) capital accumulation. Understanding the dynamics driving imperialism, and the structures and policies that ensure the continued flow of both resources and profit from certain regions and social groups to other regions and social groups is, I suggest, the study of 'ecological imperialism.'

Also, ecological imperialism and EUE may be related but they should be kept analytically distinct. EUE mainly refers to unequal flows of resources from one region or country to another; ecological imperialism refers to a political-economic arrangement that enforces identifiable structures of domination and socio-ecological degradation. Care must be taken not to reduce ecological exploitation to the unequal flows of resources as unequal biophysical flows do not always capture unequal dynamics of power *prima facie*. In order to obtain the full picture concerning ecological exploitation, we must also look at the politico-economic and social context of such flows, including flows of profit and ownership of resources, the socio-ecological *impact* of the appropriation of resources rather than just the *quantitative* amount, and the history of power struggles over resources. Ecological imperialism, as I discovered in a case study on foreign investment in Tanzania (Frame 2015), may or may not actually result in EUE measured in material flows.

Examining both the dynamics of EUE and ecological imperialism helps us grasp the role of semi-peripheral countries in the global ecology. In terms of EUE, as I discuss in the following sections, semi-peripheral countries seem to play an ambiguous role as both net suppliers and net recipients of flows of biophysical resources. However, this is not a static position. As will be seen, such countries are rapidly increasing their domestic material consumption (DMC), with increasing reliance on imported resources. Both roles, as net suppliers and net recipients, are pushing relations of ecological imperialism with their peripheral neighbors.

The Dynamics Driving Regional Imperialism: The Case of China, Malaysia, Thailand, and Vietnam in Cambodia's Land Sector

Land grabbing in Cambodia has resulted in drastic and devastating social[3] and environmental[4] consequences. It is a negative term, one used to denote a variety of dubious investment practices in the land sector, legal or illegal, on the part of foreign or domestic elites, with negative social and environmental results, such as forced land evictions, land dispossession, and deforestation. What is notable about the investment in Cambodia's land sector is the fact that the majority of investors are from neighboring East Asian or Southeast Asian countries. If we take income group rankings of the World Bank (low income, lower-middle and upper-middle income, and high income, as gross national income (GNI) per capita, fiscal year 2016) as roughly comparable to the categorizations of peripheral (low income), semi-peripheral (lower-middle and upper-middle income), and core (high income), we see in Table 4.1 that with the exception of Cambodian investors themselves, all other main investors come from Asian countries that are relatively more affluent than Cambodia.

Table 4.1 Investors in Cambodia's land sector, income profile, and land concessions

Investor country origin	Income group[a]	Number of concessions	Hectares (total)
Cambodian	Low income	112	906,161
Vietnamese	Lower-middle income	55	356,560
Chinese	Upper-middle income	42	369,107
Malaysian	Upper-middle income	12	90,844
Thai	Upper-middle income	7	59,663
Korean	High income	9	90,548
Singaporean	High income	11	137,815
Other		12	81,253
Unknown		12	28,866
Total		272	2,120,817

Income rankings based on GNI per capita, fiscal year 2016, is from World Bank Data http://data.worldbank.org/about/country-and-lending-groups: [a]Low income = $1045 or less, Lower-middle income = $1046–$4125, Upper-middle income = $4126–$12,735, High Income = $12,736 or more

Land concession information is taken from the LICADHO website, http://www.licadho-cambodia.org/land_concessions/

All concessions are economic land concessions (ELCs). ELCs are long-term leases, granted by the Cambodian government, that allow a concessionaire to clear land in order to develop industrial-scale agriculture, and can be granted for various activities including large-scale plantations, raising animals, and building factories to process agricultural products

In contrast, Cambodia and Laos, two of the lowest income countries in Southeast Asia, are not investing in land deals in other countries (Land Matrix 2017). This corresponds with the trends noted in other studies on land grabbing. Land grabbing primarily originates from three groups of countries: those in the Global North, emerging countries (e.g., Brazil, South Africa, China, India, Malaysia, and Korea), and the Gulf states. Overall, studies have shown that investment is coming from wealthier, food-importing countries with an average gross domestic product (GDP) per capita four times higher than the target countries (Oakland Institute 2011). In contrast, targeted countries are among the poorest nations in the world and tend to have high rates of hunger. Investors also tend to seek out countries with weak land institutions that also offer relatively high levels of investor protection. As food and energy security are primary drivers of investment, most investment projects are export-oriented and target high-yield, easily accessible arable land with access to water resources.

The case of Cambodia typifies 'regional ecological imperialism' marked by regional EUE in land. In Cambodia, and potentially elsewhere throughout the peripheral world, such imperialism is not only or even primarily carried out by countries in the Global North, but instead by relatively more developed countries within the same general region.

In this 'regional ecological imperialism' there is a 'pecking order,' whereby the poorest countries in a particular region are subject to land/resource grabbing, even by neighbors only somewhat better off, like Vietnam. It should be noted, however, that even though the majority of investment is regional, countries of the Global North such as Australia, the United States, Canada, the Netherlands and others have invested in smaller land deals in Cambodia.

The dynamics behind the semi-peripheral/regional imperialism observed in Cambodia are complex and arise from the East Asian and Southeast Asian countries' ambiguous 'middle' role in both the global ecology as well as their 'middle' position in the global economy. On the one hand, as China's case infamously demonstrates, competition to capture a higher share of the global surplus through rapid industrialization is intense; in order to be competitive their own domestic environments and large labor surplus are appropriated and degraded, with the costs externalized to the greater society while the benefits accrue to an elite capitalist class. On the other hand, industrialization is not the only factor that leads to the degradation of domestic semi-peripheral natures. In addition, as proponents of world-systems theory note, semi-peripheral economies contain both a mixture of core and peripheral activities. Environmental degradation also occurs as a result of the 'peripheral' economic activities they continue to engage in, as deforestation due to both domestic and foreign-owned large-scale agribusinesses and timber companies in Malaysia, Vietnam, and Thailand demonstrates. Both the industrialization and continued peripheral activities in these countries are often built upon domestic processes of accumulation by dispossession, environmental degradation, and social unrest, a dynamic that scholars have noted in other countries as well (Gellert 2015; Hall 2013). The response to these dynamics has been a geographic fix, an outflow of investment in other more peripheral countries in search of land, resources, and energy. As a consequence of their intermediate position, semi-peripheral countries are both driven to exploit peripheral environments and subjected to exploitation.

Finally, even as semi-peripheral countries engage in imperialism with their peripheral neighbors, core countries remain implicated in multiple ways. First, in many cases semi-peripheral economies operate as suppliers and subcontractors to corporations in the Global North, and many of their exports end up consumed in the Global North. Second, international financial institutions, particularly the World Bank and the International Monetary Fund, have been instrumental in promoting the industrialization and neoliberalization of semi-peripheral natures. But they have also opened up peripheral economies and enabled the appropriation of their natures. The following sections discuss these dynamics in the case of the semi-peripheral countries investing in Cambodia's land sector. If we follow the logic of (Arrighi 1990; Arrighi and Drangel 1986) and others and consider the GNI a rough indicator of a country's position in the tripartite world-system, the major investors in Cambodia's land sector, including China, Thailand, Vietnam, and Malaysia, could be considered 'semi-peripheral,' or at the least relatively better off peripheral countries struggling to obtain semi-peripheral status.

Industrialization and the Changing Biophysical Needs of the Emerging Asian Economies

As discussed above, theoretical work on EUE has been largely couched in the binary terms of core and periphery, developed or less developed, and industrialized or periphery, with core regions supposedly net importers of resources, and peripheral regions net exporters. Such binary categorizations become increasingly problematic as countries in the Global South industrialize, because they are not solely exporters of natural resources. Instead, as environmental sustainability scholars have empirically documented, as countries industrialize, resources from within national boundaries become insufficient for increasing production demands, and also consumption demands rise along with living standards (UNCTAD 2012). Significant to the issue of EUE, the physical trade balances (PTBs) of countries in the region show an increasing reliance on the imports of certain resources, especially for China. With industrialization and also urbanization, economies in this region are in the process of transitioning away from being biomass-based (agricultural) to minerals based. Their PTBs reflect this transition in increased net imports of metal ores and industrial minerals and fossil fuels. As a whole, the Asia-Pacific region is reverting toward a higher greenhouse gas emitting fuel mix. Further, the United Nations Environment Program (UNEP) (2013) report states that based on current trends, future extractive pressures on the environment will increase even more rapidly than the region's rapid rate of growth. Such trends can be contrasted with the PTB of Africa, a region where many countries are typically considered 'peripheral.' Since 1985, the aggregate PTB of all countries throughout Africa shows a rapidly increasing *net export* of the materials needed to industrialize—fossil fuels, metal ores, and minerals (UNCTAD 2012) (see Fig. 4.1a and b).

China's rapid economic growth and industrialization has driven a correspondingly rapid growth in DMC per capita (see Fig. 4.1b). In economy-wide material flows accounting 'domestic material consumption' equals domestic extraction plus imports minus exports in physical weight. DMC is used as a measure of the total amount of material directly used in the economy (excepting hidden flows). While it does not directly show EUE (let alone ecological imperialism), DMC usefully shows the aggregate amount of resources used by a country. As such, trends in DMC across time and across countries can demonstrate a country's increasing consumption of resources (including imports), and inequalities in resource consumption globally, respectively. As Fig. 4.1b indicates, China's transition away from a biomass-based to mineral-based economy has accelerated. China exceeded the world DMC average in 2003 (the average was 10 tons per capita), and by 2008 was consuming 160 percent above the world average (approximately 11 tons per capita.)

Figure 4.1a shows China's PTB. The PTB indicator (imports—exports in physical tons) has been used in a number of studies (Perez-Rincon 2006; UNEP 2011; Vallejo 2010) as an approximation of EUE. EUE has

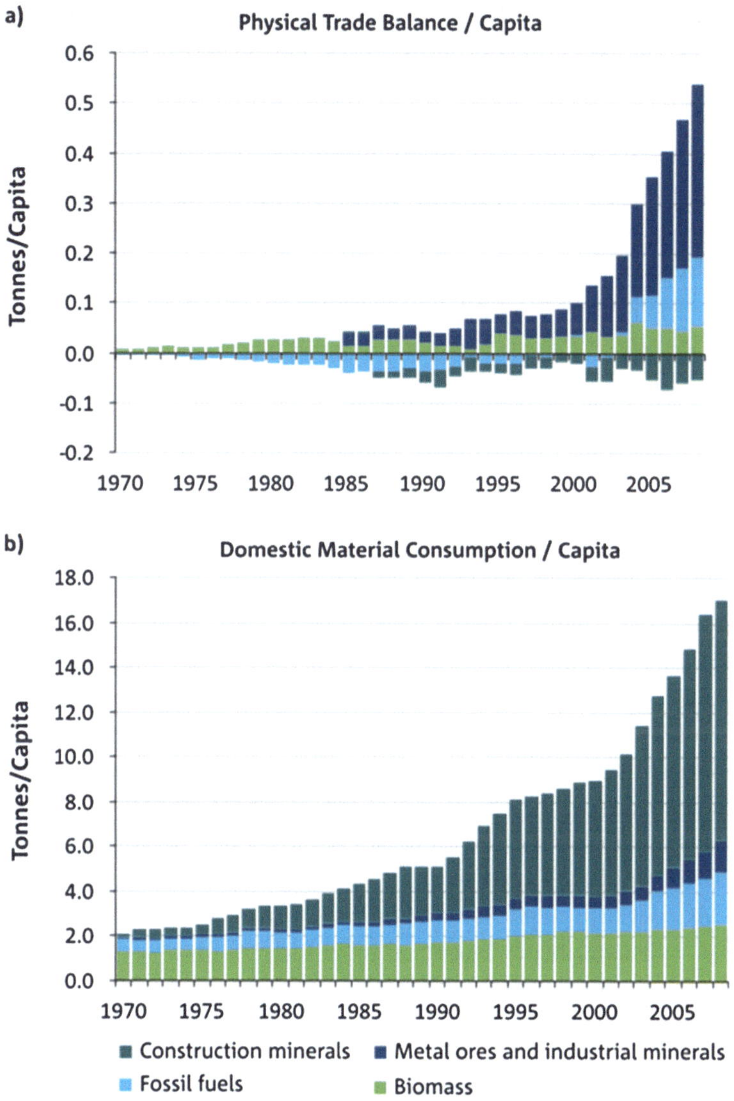

Fig. 4.1 (a) China's physical trade balance per capita and (b) domestic material consumption per capita. Source: UNEP Report (2013)

been defined, as mentioned above, as the asymmetrical transfer of resources (Hornborg 1998, 2001, 2011) from one region/country to another. The PTB indicator, unlike monetary trade indicators, shows net imports or net exports based on physical tons of material. In the above figures, the PTB shows an increasing reliance of China on imports to meet its requirements for metals, fossil fuels, and even biomass.

Malaysia (see Fig. 4.2a and b) is considered a semi-peripheral country, with the economy being pushed toward higher-technology products under the goal of becoming a fully developed economy by 2020. Malaysia's DMC per capita has increased from approximately 4 tons per capita in 1970 to approximately 17 tons per capita in 2008. The PTB reported in Fig. 4.2a shows that while overall Malaysia is still a net exporter of fossil fuels, biomass, and construction materials, this trend is decreasing, and Malaysia is growing as a net importer of metal ores and industrial minerals (UNEP 2013).

Like China and Malaysia, Thailand's (see Fig. 4.3a and b) industrialization process has resulted in a steadily increasing DMC per capita. In terms of trade, Thailand shows both an increasing reliance on the import of fossil fuels and metal ores and industrial minerals, but it is also increasingly a net exporter of biomass and construction minerals.

Vietnam (see Fig. 4.4a and b) is one of the world's fastest-growing economies. The country began the transition to a market economy in 1986. Its industrial production has led to recent annual GDP growth rates of more than 7 percent, and in 2007 the GDP grew by 17 percent. Vietnam's rapid industrialization is reflected in its equally rapid increase in DMC per capita, particularly in construction minerals. Vietnam has steadily increased its net exports of fossil fuels, but has also become a net importer of construction minerals and metal ores and industrial minerals.

Industrialization, Domestic Primitive Accumulation, and Land Grabbing as a Geographic Fix

Aside from increasing their net imports of primary products, each of the four countries is also engaged in large-scale land investments overseas.

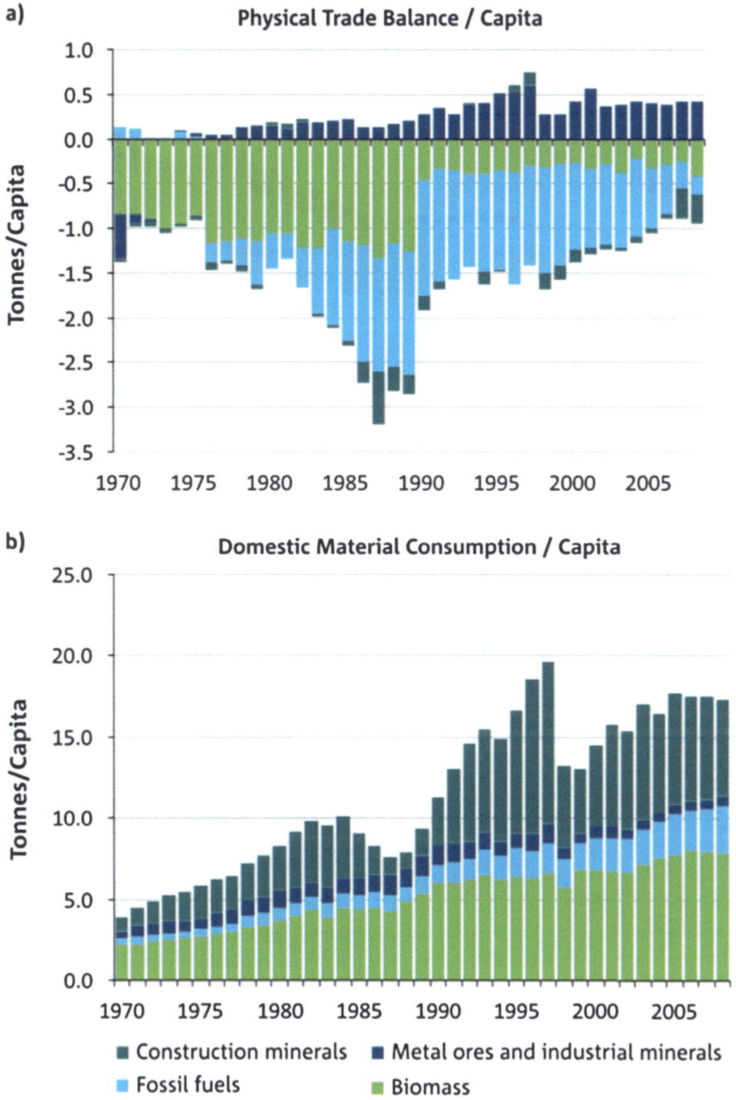

Fig. 4.2 (a) Malaysia's physical trade balance per capita and (b) domestic material consumption per capita. Source: UNEP Report (2013)

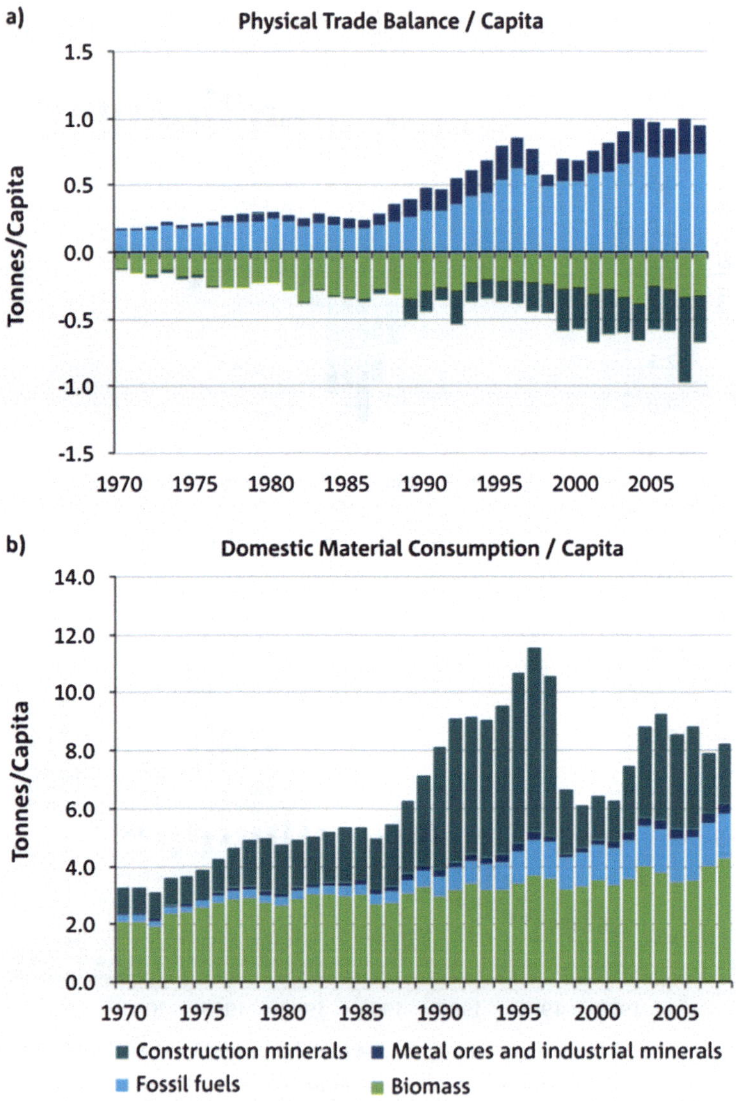

Fig. 4.3 (a) Thailand's physical trade balance per capita and (b) domestic material consumption per capita. Source: UNEP Report (2013)

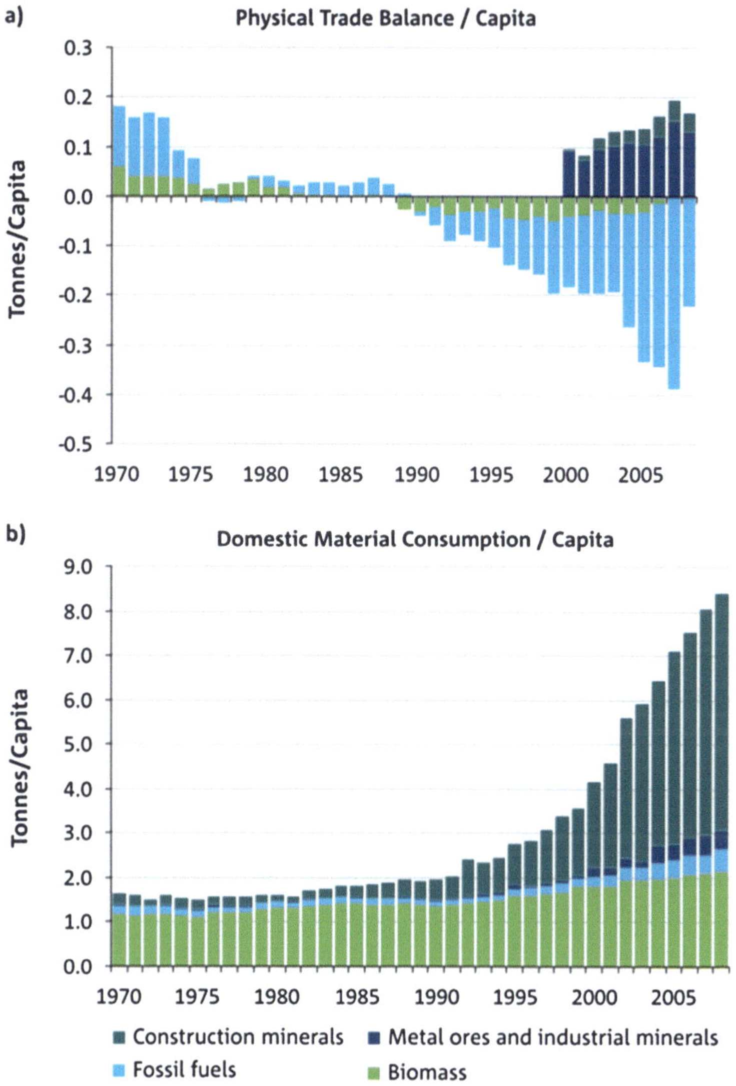

Fig. 4.4 (a) Vietnam's physical trade balance per capita and (b) domestic material consumption per capita. Source: UNEP Report (2013)

Malaysia and China are important major global players in the land grabbing trend. According to the Land Matrix (2017) data, China has invested in at least 1,380,241 hectares of land in Africa, Southeast Asia, Central Asia, South America, Central Asia, the Caribbean, and Eastern Europe. Malaysia has investments throughout Africa, South America, and Southeast Asia encompassing upward of 3,321,919 hectares. According to Land Matrix (2017) data, the vast majority of Malaysia's overseas investments are concentrated in the agricultural sector, especially palm oil but also in the forestry sector. On a lesser scale, Thailand is engaged in at least 343,513 hectares of land deals, all in other Southeast Asian countries. Vietnam land investments total 323,599 hectares in Western Africa and Southeast Asia.

With the exception of Thailand, these countries have land under foreign investment, reflecting their semi-peripheral status. The Land Matrix (2017) reports at least 262,933 hectares of Chinese soil under foreign investment, mainly from North America, Northern Europe, Australia and New Zealand, Southeast Asia, and East Asia. Inflows into Malaysia's land sector are up to 294,644 hectares, mainly from East Asia, South Asia, Australia and New Zealand, the Middle East, and Northern Europe for agricultural, industry, and forestry investments. For Vietnam, up to 351,809 hectares have been invested in by East Asia (Hong Kong and Japan) and the Middle East. It is worth noting that the Land Matrix (2017) has recorded zero land investments inflows into the United States, but net outflows of at least 8,525,138 hectares.[5] Likewise, other 'core' or high income economies such as Switzerland, the Netherlands, Luxemburg, Germany, France, the United Kingdom, Sweden, Norway, Denmark, and Finland have no recorded inflow land investments. They all, however, have invested in hundreds of thousands, if not millions, of hectares of land overseas.

Arrighi et al. (2003) note that intermediate status semi-peripheral countries are subject to intense competition in efforts to capture a greater share of the global surplus. Critical scholars have linked investments emanating from the semi-peripheral countries with rapid industrialization and domestic primitive accumulation. For example, Muldalvin (2012) argues that China's land grabs are a 'geographic fix,' driven by a development model of rapid industrialization and economic growth ultimately based on environmental destruction, social decay, and primitive accumulation. China's heavy demand for resources and energy to fuel its rapid industrialization spurred its own domestic land grab, through land and resource acquisitions by state and non-state actors. Such domestic land grabbing has led to widespread landlessness and loss of livelihoods, with an estimated 75 million landless peasants and 200 million migrants and widespread social unrest. In response to fears of declining legitimacy, Muldalvin (2012) maintains the leadership has attempted to intensify the development model of continuous high growth, bringing about further environmental and social degradation. In response to this contradiction, China is engaging in international land grabs as a 'geographic fix.' Externalizing resource demand through global sourcing and production, China is attempting to decrease domestic environmental destruction and land losses while maintaining rapid capital accumulation and continued industrialization (Muldavin 2012).

Land Grabbing and the Intermediate Position in Global Commodity Chains

The land grabbing of the semi-periphery also reflects their intermediate position in the hierarchy of global commodity chains. As Muldavin (2012) argues, while China is a competitive concern for core economies, many global corporations are integrated in China's industrial platform and overseas subcontracting relations. China thus plays an important role for such corporations by opening up new areas for resource flows, keeping global prices down for some commodities, and facilitating super profits for such global corporations. Transnational corporations, Muldavin (2012) insists, have in essence subcontracted China to extract resources from Africa, Asia, and Latin America, further obscuring the linkages between the social and ecological costs of extraction and production, and consumption. For example, China's large demand for timber resources has led to decimation of its own forests, both as a consequence of increasing domestic demand and internal consumption, but also for the global subcontracting factories on its industrial platform supplying IKEA, Walmart, and others. In response, China has shifted timber sourcing to

Siberia, Southeast Asia, Central Africa, and Latin America, once again displacing some of the impacts through a geographic fix. Similar comparable trends could be found in the case of mineral extraction and processing, lands for agribusiness, large-scale infrastructure projects, and so on. Such trends are true for the other countries as well.

In terms of the implication for core economies in semi-peripheral imperialism, Muldavin (2012) also points out that it was the very policies of such institutions as the World Bank that set the stage for China's domestic primitive accumulation and subsequent going-out investments. From 1981 the World Bank pushed privatization and market-oriented policies in all sectors of China's economy, including international support for infrastructure development, the dams, roads, rail, canals, pipelines, transmission lines, and so on. The International Finance Corporation (IFC) and the Export-Import Banks (ExIm) encouraged Foreign Direct Investment (FDI) in China, and pushed the idea of China as a global subcontracting platform. The associated infrastructure development has been an important factor behind the displacement, land loss, and dispossession.

Land Grabbing and the Peripheral Economic Activities in the Semi-Periphery

Peripheral economic activities within the semi-periphery are also major factors driving, once again, domestic primitive accumulation and socio-ecological degradation and the geographic fix manifested in overseas land grabbing. Some of the peripheral economic activities have led to the growth of large agribusinesses and timber tycoons that are implicated in both domestic land grabs and land grabs in other countries. While Malaysia exports semiconductors, electronic equipment, solar panels, and textiles, it is also a major exporter of tropical hardwood, palm oil, wood and wood products, rubber, petroleum and liquefied natural gas, and chemicals. Thailand's export economy is concentrated in the automobile, petrochemical, and electronics sectors, but it also includes rice, rubber, and fishery products. Though industrializing rapidly, Vietnam exports clothes, shoes, electronics, seafood, crude oil, rice, coffee, wooden products, and machinery. The exports end up in core or other semi-peripheral countries. The top export partners of these countries are other Asian countries, the United States, and Australia (Index Mundi 2017 Website).

One of the most significant environmental consequences of continued peripheral activities in these countries is deforestation. Deforestation, and the attendant loss of biodiversity and other socio-ecological consequences, is a major environmental issue in Southeast Asia as a whole. Malaysia has one of the world's highest rates of deforestation; the major agents of deforestation include commercial logging, commercial oil palm and other tree planting, agribusiness expansion, large dams, extractive industries such as mining, infrastructure and urban development projects, consumer demands for logs and palm oil, particularly among food producers and the biodiesel industry. As in China and Cambodia, deforestation in Malaysia has been executed through appropriation and primitive accumulation. Activists have accused the country's powerful political and economic elites of plundering the rich resources and wealth of the country for quick profits. Community forests are being cleared, with harmful impacts on communities' access to forest resources for livelihoods and food security and consequent intensification of livelihood hardship and poverty. In the last few decades, Malaysia's once abundant and biodiverse rainforests have been rapidly destroyed or damaged (Yang 2014). Malaysia lost 4.7 million hectares of tree cover during 2000–2012, putting it among the top 10 countries for the percentage of tree cover lost (World Resources Institute 2014). Now, with most of the economically attractive forest areas for timber extraction gone, the timber industry is faltering and being replaced by large-scale palm oil plantation development and industrial tree plantations (Yang 2014).

Vietnam's rapid economic growth has increased environmental degradation, particularly deforestation. More than 60 percent of the country was originally covered by forests, but war, logging, population growth, energy production, and land conversion have reduced forest cover to 10–30 percent in the country. Similarly, forests covered 53 percent of Thailand's land area in 1961, but now this figure is 28 percent, resulting from agricultural activity, expanding urban areas, and both legal and illegal logging (Krechowicz and Fernando 2009). In response to such domestic degradation, the large agribusinesses located in the semi-periphery are shifting their quest for resources externally.

The Effects of Regional Ecological Imperialism in Cambodia's Land Sector

As a geographic response to the dynamics discussed above, Thai, Malaysian, Chinese and Vietnamese companies are investing in peripheral countries like Cambodia in a manner that mirrors the trends of primitive accumulation, dispossession, and environmental degradation in their own countries. Cambodia's land grabbing hinges on corruption, lack of democracy as well as economic policies which favor investors over environmental and social concerns. In the last decade, the government has been involved in the forcible displacement of hundreds of thousands of Cambodians from their homes and farmlands, and many more threatened with displacement. Since 2000, Cambodia's national human rights organization the Cambodian League for the Promotion and Defense of Human Rights (LICADHO) has collected data documenting that over half a million people have been affected by land conflicts. LICADHO has copious documentation of the civil unrest, human rights abuses, and corruption underlying this trend, a trend enabled by several key actors: the state and military, investors both foreign and domestic, and also international financial organizations, in particular the World Bank. In the majority of cases, the Cambodian state has played a critical role through the granting of land concessions and the use of state forces to intimidate people, remove them from their land, and destroy homes. While legally required to consult with local communities and undertake environmental impact assessments before granting land as an economic concession, such procedures are routinely ignored. In a country where democracy and the rule of law are thinly executed, activists are often subject to threats and intimidation from the government and the frequent use of force and the law against them. Many of the displaced have lost their land, homes, and often their livelihoods. With at least 500,000 Cambodians affected by state-involved conflict over land, the sheer number of anecdotal stories of land grabbing is overwhelming.[6]

It should be noted that, like other peripheral countries, international financial institutions were central in pushing the legal framework that has enabled the land grab in Cambodia. For Cambodia, it was the IFC, the private sector arm of the World Bank, that provided the financial and technical support to create Cambodia's new investment climate. In 2003, the Cambodian government amended its 1994 Investment Law, ensuring a wave of deregulation meant to encourage FDI inflows. These reforms included cuts on certain import duties, renewable land leases of up to 99 years, and no price controls on goods or services produced by investors. This triggered sharp increases in the FDI, including agriculture and natural resource extraction (Oakland Institute Report 2013). Like other peripheral countries that have undergone similar foreign investment regulatory changes, land grabbing is now justified by country elites under the auspices of economic 'development.'

Environmentally, land grabbing in Cambodia has been strongly linked to deforestation (LICADHO). Nearly half of the areas where concessions were granted between 2000 and 2012 was forested in 2000, an area that had represented 12.4 percent of forest land cover in Cambodia. With concessions, the annual rate of forest loss was between 29 percent and 105 percent higher than in comparable land areas outside concessions (Davis et al. 2015). Land concessions, many of them agricultural and which are supposed to occur on degraded lands, are being used to clear-cut large areas of Cambodia's oldest, most valuable, and most biodiverse remaining forests—despite the fact that Cambodia officially ended logging concessions in 2001.

The vast majority of land grabbing in Cambodia is for agricultural purposes (see Table 4.2), about 1,295,778 hectares are devoted to the cultivation of rubber, sugar, paper pulp, cassava, and palm oil. Sugar production claims 79,172 hectares of Cambodia's land concessions, mainly from Chinese and Thai investors. According to a LICADHO (2014) report, many Cambodian sugar plantations are built on land violently wrested from poor farmers, resulting in civil unrest and conflict. Yet as discussed above, core countries remain implicated, even if the actions are being carried out by investors from the semi-periphery. Since 2009, an increasing number of Economic Land Concessions in Cambodia have been used to produce raw sugar. This trend is a result of the European Union's (EU) Generalized

Table 4.2 Land concession area in Cambodia by country of investor (China, Vietnam, Malaysia, and Thailand) and crop[a]

Country	Rubber	Rubber (mixed)	Sugar	Paper pulp	Cassava	Palm oil	Other	Unknown
China	30,637	89,882	60,072	72,115	5,789	0	54,954	55,658
Vietnam	312,687	8,520	0	0	0	25,973	0	9,380
Malaysia	24,836	7,800	0	0	7,955	6,718	0	43,535
Thailand	15,450	0	19,100	0	0	8,120	0	16,993

[a]All concessions listed are 'Economic Land Concessions' in units of hectares, taken from the LICADHO website

System of Preferences (GSP) preferential trade scheme, which allows Cambodia to export sugar to the EU duty free. Approximately 92 percent of sugar exports went to the EU. Further, Thai sugar giant Mitr Phol is one of Coca Cola's top three international suppliers, and was accused of land grabbing in Cambodia's Oddar Meanchey province before these plans were abandoned amid social unrest in late 2014 (Peter and Pheap 2016).

The production of palm oil is behind 40,811 hectares of land concessions in Cambodia, stemming from Malaysian, Thai, and Vietnamese companies. Like sugar, the expanding investment in ELCs for producing palm oil stems not just from developing country demand, but from an increasing global demand. Palm oil production is a notoriously environmentally destructive activity responsible for high levels of deforestation throughout Southeast Asia. Malaysia and Indonesia alone produce over 80 percent of internationally traded crude palm oil (Sokhannaro 2012). Thailand and Papua New Guinea are now also experiencing a rush to expand the crop and there are initiatives to further develop the crop in Cambodia, Vietnam, and the Philippines. Palm oil production represents a salient example of a geographic expansion of peripheral activities within semi-peripheral states to their less-developed neighbors.

Rubber production, dominating the largest proportion of land concessions in Cambodia at 489,812 hectares, is another such example. One consequence of the rapid deforestation in the Southeast Asian economies is growth in land grabs that are thinly veiled, sometimes illegal, logging operations in poorer countries. For example, in Cambodia two giant Vietnamese companies, Hoang Anh Gia Lai (HAGL) and the Vietnam

Rubber Group (VRG) and their subsidiaries, have carried out forced land grabs without compensation, and satellite images show they are logging illegally in protected forests. Violence, harassment, and forced evictions have been regularly reported in interviews with local groups. Thousands of households have been deprived of their livelihoods from forest products or subsistence agriculture. As noted, international financial institutions such as the World Bank have been instrumental in opening up peripheral areas to FDI through encouraging deregulation in the natural resource and agricultural sectors. Beyond this, they have also actively supported private firms involved in such acquisitions. HAGL and VRG have both received substantial foreign investments from Deutsche Bank and the IFC, either directly or via intermediary funds (Global Witness Report 2013).

Conclusion

This chapter has attempted to highlight several trends important to the study of EUE. Overall, it sought to reiterate how the intermediate position of semi-peripheral Asian countries in the global economy leads to complex dynamics that enforce a rather ambiguous position as both exploiter and exploited. While this has been analyzed in regard to the global hierarchy of wealth, in a world of ecological crises and dwindling resources, this intermediate position is also significant as it has pushed both the rapid exhaustion of their own natures, while simultaneously instigating a relationship of ecological imperialism with their peripheral neighbors. This marks a major change in the history of imperialism. Under colonialism, imperialism was driven by the colonial powers, with countries throughout Asia, Africa, and Latin America 'imperialized,' or, in world-systems terminology, drawn into the international division of labor for the purpose of providing cheap, easily accessible raw materials for Europe's industrial processes. Now, for all the dynamics this chapter discusses, semi-peripheral in the Global South states such as the East Asian emerging economies are engaging in ecological imperialism in the periphery, opening up peripheral resources to exploitation, using peripheral resources to meet their metabolic needs, and drawing countries like Cambodia deeper into regional economic integration. Imperialism has always been predicated upon the appropriation of peripheral natures. However, the uneven but nonetheless global saturation of the capitalist mode of production is driving new dynamics of appropriation from not just core, but newly 'emerging' industrializing centers, even as their own natures are exploited in the intense competition to industrialize and serve as subcontractors to companies in the Global North.

While this chapter articulates some of the central dynamics behind semi-peripheral appropriation of peripheral natures, the case study is small. Large-n, empirical research examining the extent of EUE between the semi-periphery and the periphery would be illuminating, as would studies that differentiate between semi-peripheral countries. These are important areas for future research.

Notes

1. Such broad definitions also do not necessarily exclude the possibility of EUE within countries. However, like the case of EUE between the semi-periphery and periphery, such a significant form EUE needs to be explicitly included and clarified theoretically.
2. See, for example, 'China's Influence in Africa: Implications for the US' http://www.heritage.org/research/reports/2006/02/chinas-influence-in-africa-implications-for-the-united-states.
3. LICADHO's website has an entire section devoted to articles, statements, and videos on the social and environmental costs of land evictions in Cambodia. See for example https://www.licadho-cambodia.org/pressrelease.php?perm=342.
4. See LICADHO's website https://www.licadho-cambodia.org/land_concessions/ for an interactive map of Cambodia's deforestation and its relation to land concessions.
5. However, in recent years some foreign investment have been occurring in US land. See http://www.takepart.com/article/2014/02/12/land-grabs-at-home.
6. The LICADHO website holds numerous reports on the social, political, and environmental issues involved with the land grabbing phenomenon in Cambodia, http://www.licadho-cambodia.org/.

References

Arrighi, Giovanni. 1990. "The Developmentalist Illusion: A Reconceptualization of the Semi-Periphery." Pp. 11–42 in *Semi-Peripheral States in the World-Economy*, edited by W. G. Martin. Westport, CT: Greenwood Press.

Arrighi, Giovanni and Jessica Drangel. 1986. "The Stratification of the World-Economy: An Exploration of the Semiperipheral Zone." *Review (Journal of the Fernand Braudel Center)* X(1):9–74.

Arrighi, Giovanni, Beverly Silver, and Benjamin Brewer. 2003. "Industrial Convergence, Globalization and the Persistence of the North-South Divide." *Studies in Comparative International Development* 38(1):3–31.

Borras, Saturnino, Jennifer Franco, Sergio Gomez, Christobal Kay, and Max Spoor. 2012. "Land Grabbing: Latin America and the Caribbean." *The Journal of Peasant Studies* 39(3–4):845–872.

Bunker, Stephen G. 1985. *Underdeveloping the Amazon: Extraction, Unequal Exchange, and the Failure of the Modern State*. Chicago, IL: University of Chicago Press.

Chase-Dunn, Christopher. 2005. "Contemporary Semi-Peripheral Development: The Regimes and the Movements." Institute for Research on World-Systems, University of California-Riverside. Retrieved August 5, 2016 at https://irows.ucr.edu/papers/irows78/irows78.htm.

Chase-Dunn, Christopher and Thomas Hall. 1997. *Rise and Demise: Comparing World-Systems*. Boulder, CO: Westview Press.

Ciccantell, Paul and David Smith. 2009. "Rethinking Global Commodity Chains: Integrating Extraction, Transport, and Manufacturing." *International Journal of Comparative Sociology* 50(3–4):361–384.

Clark, Brett and John B. Foster. 2009. "Ecological Imperialism and the Global Metabolic Rift: Unequal Exchange and the Guano/Nitrates Trade." *International Journal of Comparative Sociology* 50(3–4):311–334.

Davis, Kyle Frankel, Kailiang Yu, Maria Cristina Rulli, Lonn Pichdara, and Paolo D'Odorico. 2015. "Accelerated Deforestation Driven by Large-Scale Land Acquisitions in Cambodia." *Nature Geoscience* 8:772–775.

Foster, John B., Brett Clark, and Richard York. 2011. *The Ecological Rift: Capitalism's War on the Earth*. New York: Monthly Review Press.

Frame, Mariko. 2014. *Foreign Investment in African Resources: The Ecological Aspect of Imperialism and Unequal Exchange*. Unpublished Ph.D. Dissertation, Korbel School of International Studies, University of Denver, Denver, CO.

———. 2015. "The Neoliberalization of (African) Natures as the Current Phase of Ecological Imperialism." *Capitalism, Nature Socialism* 27(1):87–105.

———. 2016. "Economic Integration in Tanzania (1970–2011): A Biophysical Assessment." *Journal of Industrial Ecology* 20:1083–1093.

Gellert, Paul K. 2015. "Palm Oil Expansion in Indonesia: Land Grabbing as Accumulation by Dispossession." Pp. 65–99 in *States and Citizens: Accommodation, Facilitation and Resistance to Globalization*, edited by Jon Shefner. New York: Emerald Group.

Global Witness. 2013. "Rubber Barons." Retrieved February 15, 2016 at https://www.globalwitness.org/en/campaigns/land-deals/rubberbarons/.

Hall, Derek. 2011. "Land Grabs, Land Control, and Southeast Asian Crop Booms." *The Journal of Peasant Studies* 38(4):837–857.

———. 2013. "Primitive Accumulation, Accumulation by Dispossession and the Global Land Grab." *Third World Quarterly* 34(9):1582–1604.

Hornborg, Alf. 1998. "Ecosystems and World-Systems: Accumulation as an Ecological Process." *Journal of World-Systems Research* 4(2):169–177.

———. 2001. *The Power of the Machine: Global Inequalities of Economy, Technology, and Environment*. Walnut Creek, CA: AltaMira Press.

———. 2011. *Global Ecology and Unequal Exchange: Fetishism in a Zero-sum World*. New York: Routledge.

Index Mundi. 2017. *Country Facts*. Available at https://www.indexmundi.com.

Jorgenson, Andrew. 2016. "Environment, Development and Ecologically Unequal Exchange." *Sustainability* 8:227.

Klare, Michael. 2012. *The Race for What's Left: The Global Scramble for the World's Last Resources*. New York: Picador.

Krechowicz, Dana and Hiranya Fernando. 2009. "Emerging Risks and Impacts of Key Environmental Trends in Emerging Asia." World Resources Institute. Retrieved February 17, 2016 at http://www.wri.org/publication/emerging-risk.

Land Matrix. 2017. Database available at http://landmatrix.org.

Li, Minqi. 2003. "The Rise of China and the Demise of the Capitalist World-Economy: Exploring Historical Possibilities in the 21st Century." Retrieved December 12, 2015 at http://www.networkideas.org/featart/mar2004/rise_of_china.pdf.

LICADHO. 2014. "Year 2013 in Review: Land, a Country in Crisis." Retrieved January 15, 2016 at http://www.licadho-cambodia.org/articles/20140321/142/index.html.

Magdoff, Harry. 2003. *Imperialism Without Colonies*. New York: Monthly Review Press.

———. 2013. "Twenty-First-Century Land Grabs: Accumulation by Agricultural Dispossession." *Monthly Review* 45(6):1–18. Retrieved August 7, 2016 at http://monthlyreview.org/2013/11/01/twenty-first-century-land-grabs/.

Moore, Jason W. 2001. "Marx's Ecology and the Environmental History of World Capitalism." *Capitalism, Nature Socialism* 23(3):134–139.

———. 2003. "The Modern World-System as Environmental History? Ecology and the Rise of Capitalism." *Theory and Society* 32(3):307–377.

———. 2011. "Transcending the Metabolic Rift: A Theory of Crises in the Capitalist World-Ecology." *The Journal of Peasant Studies* 38(1):1–46.

———. 2012. "Cheap Food and Bad Money: Food, Frontiers, and Financialization in the Rise and Demise of Neoliberalism." *Review (Journal of the Fernand Braudel Center)* 33(2–3):225–261.

Muldavin, Joshua. 2012. "Land from the Tiller: China's Role in Global Processes of Land Dispossession." Conference paper, Land Deal Politics Initiative, Institute of Social Sciences, Cornell University, Ithaca, NY.

Oakland Institute. 2009. "The Great Land Grab Rush for the World's Farmland Threatens Food Security for the Poor." Retrieved August 5, 2016 at http://www.oaklandinstitute.org/sites/oaklandinstitute.org/files/LandGrab_final_web.pdf.

———. 2011. *Understanding Land Investment Deals in Africa: Country Report Tanzania*. Oakland, CA: The Oakland Institute.

———. 2013. "World Bank's Bad Business in Cambodia." Retrieved March 12, 2013 at http://www.oaklandinstitute.org/sites/oaklandinstitute.org/files/OurBiz_Fact_Sheet_Cambodia.pdf.

Perez-Rincon, Mario Alejandro. 2006. "Colombian International Trade from a Physical Perspective: Towards an Ecological Prebisch Thesis." *Ecological Economics* 59:519–529.

Peter, Zsombor and Aun Pheap. 2016. "Thai Firm: Sugar Farm Damage Assessment is Almost Finished." *Farmlandgrab.org*. Retrieved June 3, 2016 at http://www.farmlandgrab.org/post/view/25763-thai-firm-sugar-farm-damage-assessment-is-almost-finished.

Rice, James. 2009. "The Transnationalization of Production and Uneven Environmental Degradation and Change in the World-Economy." *International Journal of Comparative Sociology* 50(3–4):215–236.

Ruvalcaba, Daniel Efrén Morales. 2013. "Inside the BRIC: Analysis of the Semi-Peripheral Character of Brazil, Russia, India, and China." *Austral: Brazilian Journal of Strategy & International Relations* 2(4):141–173.

Singh, Simron Jit and Nina Eisenmenger. 2010. "How Unequal is International Trade? An Ecological Perspective Using Material Flow Accounting." *Journal fur Entwicklungspolitik* XXVI:57–88.

Smith, David A. and Matthew C. Mahutga. 2009. "Trading Up the Commodity Chain? The Impact of Extractive and Labor-Intensive Manufacturing Trade on World-Systems Inequalities." Pp. 63–82 in *Frontiers of Commodity Chain Research*, edited by Jennifer Bair. Palo Alto, CA: Stanford University Press.

Sokhannaro, H. E. P. 2012. Retrieved July 5, 2016 at http://www.forestpeoples.org/sites/fpp/files/publication/2011/11/chapter-2-palm-oil development-cambodia.pdf.

United Nations. 2012. *Economic Development in Africa Report 2012: Structural Transformation and Sustainable Development in Africa*. New York: United Nations.

United Nations Environment Programme (UNEP). 2011. *Decoupling Natural Resource Use and Environmental Impacts from Economic Growth*. Paris: United Nations Environment Programme.

———. 2013. *Resource Trends in Material Productivity in Asia and the Pacific*. Retrieved May, 2014 at http://www.unep.org/pdf/RecentTrendsAP%28FinalFeb2013%29.pdf.

Vallejo, M. 2010. "Biophysical Structure of the Ecuadorian Economy, Foreign Trade, and Policy Implication." *Ecological Economics* 70:159–169.

Visser, Oane, Natalia Mamonova, and Max Spoor. 2012. "Oligarchs, Megafarms, and Land Reserves: Understanding Land Grabbing in Russia." *The Journal of Peasant Studies* 39(3–4):899–931.

Wallerstein, Immanuel. 1974. *The Modern World-System I: Capitalist Agriculture and the Origins of the European World-Economy in the Sixteenth Century*. New York: Academic Press.

World Resources Institute (WRI). 2014. "Five Overlooked Deforestation Hotspots." Retrieved August 5, 2016 at http://www.wri.org/blog/2014/03/5-overlooked-deforestation-hotspots.

Yang, Carol. 2014. "Deforestation Drivers and Human Rights in Malaysia." Forest Peoples Program. Retrieved January 29, 2016 at (https://www.google.com/search?q=A+National+Review+and+Two+Sub-Regional+Case+Studies.%E2%80%9D++Forest+Peoples+Program.&ie=utf-8&oe=utf-8&client=firefox-b-ab#q=forest+peoples+programme+malaysia+palm+oil)

5

Bunker's Ecologically Unequal Exchange, Foster's Metabolic Rift, and Moore's World-Ecology: Distinctions With or Without a Difference?

Paul K. Gellert

Introduction: Understanding and Changing Our Socionatural World

Currently, various critical sociologists, geographers, environmentalists, ecological economists, and others are facing the daunting tasks of how to both understand the world in which we live and change it. In this formulation, I intentionally echo Marx's thesis XI formulation because the most critical social thinkers, especially ecosocialists, are likewise engaging with Marx's work among others and its legacy for contemporary thought. In so doing, they are re-evaluating ontologies and epistemologies of what is increasingly described by terms such as socionature (Gellert 2005; Swyngedouw 1999); ecological rift, or social metabolism and the variant socioecological relations (Foster, Clark, and York 2010; Longo, Clausen, and Clark 2015); and world-ecology (Moore 2015a). Other related terms include critical hybridities (White, Rudy, and Gareau 2015), as well as the coevolution of socioecological systems (Kallis and Norgaard 2010;

P. K. Gellert (✉)
University of Tennessee, Knoxville, TN, USA
e-mail: pgellert@utk.edu

R. S. Frey et al. (eds.), *Ecologically Unequal Exchange*,
https://doi.org/10.1007/978-3-319-89740-0_5

Norgaard 1994) and coupled human and natural systems (Liu et al. 2007).

Three major works stand out among efforts by sociologists to understand, research, and theorize the dialectical interpenetration of "nature" and "society" and the ways in which such "socionatures" are shaped by capitalism. Moreover, each of these scholars contributes to our understanding of ecologically unequal exchange (EUE) built on the basis of unequal power relations undergirding imperialism, capitalist exploitation, and uneven development. Stephen G. Bunker (1985) introduced "ecologically unequal exchange" in *Underdeveloping the Amazon: Extraction, Unequal Exchange, and the Failure of the Modern State.* This work is widely cited as foundational to later studies of EUE due to its theoretical development of EUE and application to the Amazon basin across five centuries of underdevelopment. However, too often his work has been reduced to an obligatory but gratuitous citation. John Bellamy Foster (2000) uncovered and expanded Marx's concept of "metabolic rift" in *Marx's Ecology: Materialism and Nature* to illuminate both how Marx was not naïve and Promethean in his understanding of human relations with(in) the environment and the costs of industrialization. The metabolic rift is expressed both in a town/country divide and in the "robbing" of land, resources, and the soil at the global level. Most recently, Jason W. Moore (2015a) has critiqued the prevailing dualism of environmental sociology and developed an alternative, nondualist, "world-ecology" perspective in *Capitalism in the Web of Life: Ecology and the Accumulation of Capital.*[1] All three sociologists[2] address the unjust manner in which dominant actors in the capitalist world-system simultaneously exploit labor and nonhuman or biophysical nature while undermining sustainability. In fact, harkening back to an earlier conceptualization, one can argue that they are each addressing what James O'Connor (1988) called the "conditions of production" that, he argued, were threatened by the "second contradiction of capitalism."

Despite this unity, a polemical debate between the ecological Marxism of John Bellamy Foster and the world-ecology of Jason W. Moore erupted from the depths of what had (arguably) been a "non-debate" into the open air (Arrighi 1998; Gellert, Frey, and Dahms 2017). Sadly, the "non-debate" has been rather one-sided. In several publications, Moore (2011a,

2014, 2017a) criticizes Foster's unpacking and application of Marx's concept of the "metabolic rift" as a Cartesian dualist formulation that separates Society from Nature by frequently analyzing the impact of the former on the putatively separate latter (see also Schneider and McMichael 2010). Moore dubs this framing "Green Arithmetic." In a kind of non-response, Foster (2016) labeled Moore a "Latourian Marxist" adding, in a more vicious personal attack that Moore, "I am sorry to say, has moved to the other side, and now stands opposed to the ecosocialist movement and socialism (even radicalism) as a whole" (Angus 2016). It is difficult not to be bewildered by the sectarian marking of "sides" and the venom.[3]

Despite their differences, there are significant areas of overlap that are being overshadowed by the recent personal and sectarian venom which leads me to conclude that the insights of Bunker were not only foundational to the EUE perspective but also stand the test of time.[4] Moreover, in his mostly overlooked posthumously published book on Andean irrigation practices, Bunker (2006) offered powerful insights into what critical sociologists and geographers have taken to calling "embodied" material practices (Bakker and Bridge 2006; O'Hearn 2007; Salleh 2005). These practices illuminate possible socionatural futures that are more just and sustainable but just as difficult to achieve as the visions of Foster and Moore.

Given the three thinkers shared interest in integrating environment into a sociological analysis of the social and environmental or socionatural inequalities produced in the world and capitalism as the root of these inequalities, the disputes overshadow their commonalities. So, what are the real distinctions between Foster's and Moore's perspective? And how does Bunker's earlier contribution intersect with their view(s)? This chapter explores whether and to what extent these three scholars offer distinct ontologies, methodologies, analyses of the past, and agendas for praxis in the future. The chapter begins with a summary of the areas of agreement among the three by unpacking key points in their major works. Second, I briefly return to the Foster-Moore "rift" in order to set the context for a more measured evaluation of differences. Following that, I review the differences between their perspectives with respect to four areas: ontology, methodology, and crisis. I end with their perspectives on possible futures that might overturn the current unsustainable situation.

Foundational Agreements on the Ecological Left

Before carefully unpacking the real distinctions between the contributions of Bunker, Foster, and Moore—or of EUE, metabolic rift, and world-ecology—I begin with a brief summary of their foundational areas of agreement. It is worth reiterating the ways in which all three scholars contribute to a critically minded sociological and ecological critique of modernity and capitalism.

First, in addressing the problems facing humanity at present, they agree (if not fully as discussed below) on the excessive human degradation of the environment, especially the nonhuman parts because these parts are (still) often overlooked by social analysis (see, e.g., York and Dunlap 2012). Bunker's foundational contribution on EUE calls special attention to the degradation occurring in the process of underdeveloping extractive peripheries such as the Amazon. As he began his book (Bunker 1985:20; see Bunker, page 13, this volume), "The development or underdevelopment of any region results from the organization, coordination, and use of human and nonhuman energies and from the distribution of resources derived from and transformed in that environment or traded for resources derived from or transformed in other regions." The appearance of economic growth in the Amazon, he argued, was illusory "because it depends on the depletion of natural resources, on the disruption of the surrounding natural environment, and on the dislocation of human environments" (Bunker 1985:82).

Foster, too, is highly concerned about the environmental degradation that has occurred in particular metabolic rifts, such as that associated with soil nutrient depletion in England and the substitution of extracting Peruvian guano in the nineteenth century (Clark and Foster 2012). More generally, Foster (2009) relies on the best available natural science to argue that we are reaching a series of "tipping points" in terms of species extinction, air and water pollution, and especially climate change, a systemic shift for which he has more recently adopted the label of the "great Climacteric" (Foster 2016). Foster's ongoing investigations of the ecological thinking within Marx's nineteenth-century analysis of capitalism show that the ecological contradictions of capitalism expressed by the metabolic rift were already understood by Marx quite early. In other words, the early Marxist critics, who Foster (2016; Foster and Clark 2016) calls "first wave ecosocialists," were incorrect in their evaluation of Marx as harboring Promethean faith in the power of productive industry and ignoring ecology.

Moore is also concerned about degradation, which he prefers to examine as the "exhaustion" of frontiers. This concern underlies his world-ecology and is evident despite Foster and Clark's (2016) worry that Moore and other "production of nature" scholars more generally do not recognize the materiality of nature. The exhaustion of frontiers occurs in specific locales, such as in the first sugar producing areas of Madeira where "relations between deforestation, soil fertility, and faltering labor productivity in agriculture [were] decisive to sugar's rapid decline" (Moore 2010a:2, 2009).[5] He acknowledges the reality of scientific evidence, writing, for example, "the link between global warming, drought frequency, and global aridity is well established" (Moore 2015b:30). More generally, for Moore (2016:1), recent work on the Anthropocene embarks from a "reality" that is "quite real" of "multiple 'planetary boundaries' being crossed." Admittedly, however, there is a difference because while Foster (2016:394) emphasizes that "all major measurements of biological and social change are shown to follow a hockey-stick pattern, including the well-known increase in carbon dioxide emissions," Moore is concerned with transcending dualist, "Green Arithmetic" wherein "Capitalism plus Nature = Catastrophe" through a consideration of "something possibly more significant than the 'degradation' of nature" (Moore 2016:5). While somewhat caustically commenting that "depletion is real enough," (Moore 2015a:105), he recognizes, "Global warming poses a fundamental threat … to humanity. … [Climate change] is reinforcing tendencies—such as aquifer depletion" (Moore 2015a:290).

Second, they are all concerned about and critical of the role of capitalism in engendering the (socio)environmental degradations that abound. It is fair to argue that they all agree with Bunker and Ciccantell's

(2005:xiii) synthetic statement, "Capitalism requires territorial and spatial expansion." For Bunker, the underdevelopment of the Amazon must be understood as occurring due to "penetration" of the region by external European capitalists interested in commodifying parts of nature and who explored, attacked, attempted to control the rivers, conquered and enslaved Indians and imported indentured laborers to do so. All of this, Bunker emphasizes, was "extended by locally dominant groups who have had power to transform the environment of others" (Bunker 1985:15). For Foster, the key to understanding the relationship between capitalism and nature is the "metabolic rift" (Foster 1999, 2000; Foster et al. 2010). He frequently quotes Marx's identification of the "irreparable rift in the interdependent process of social metabolism, a metabolism prescribed by the natural laws of life itself" to argue that "capitalism as a global system has created specific environmental problems in the modern era by transgressing the universal metabolism of nature" (Foster and Clark 2016). Moore, while opening up the way humans continue to exist in varying relations or "bundles" with(in) nature, also centers his critique on capitalism. Like Bunker and world-systems scholarship more generally, he sees the reshaping of our socionatural relations within a long five centuries history of capitalism, rather than the more recent industrial revolution of the nineteenth century. Yet, he argues that Bunker is not systematic enough in that critique. For Moore (2017b:597), "Capitalism and its driving relations have indeed directed horrific violence towards human and extra-human life. I would go so far as to say that an unusual combination of productive and necrotic violence defines capitalism."

Two other interrelated points unite them. They disagree with the ideas of technological optimists and development economists that technology or other notions of sustainability such as "green growth" will save us. None of them are accepting the Breakthrough Institute's technophilic optimism that asks us to "embrace an optimistic view toward human capacities and the future" (Ecomodernist Manifesto 2015:31) and "decoupling" from environmental degradation via technological progress and modernization. While Bunker, Foster, and Moore respect human ingenuity, they reject the notion that technologies, particularly if disconnected from social relations, offer any sort of "solution" to the limits on human existence of problems like global climate change or more small-scale environmental degradation. Foster et al. (2010:89) refer to it as "Capitalism in Wonderland." Bunker (1985:238) concludes emphatically, "None of the conventional prescriptions for development" will reverse the damage and improve the Amazonian trajectory. Writing with Ciccantell (Bunker and Ciccantell 2005), he elaborates how technologies alleviate the rising costs and depletion dynamics of commodity extraction in particular locations but simultaneously expand "diseconomies of space" as familiar raw materials become scarcer, further, and deeper from capital's grasp.

Due to the Jevons' paradox of increased efficiency resulting in increased throughput of raw materials, which Bunker and Ciccantell (2003, 2005) characterized as typical of rising hegemons in their creation of "generative sectors," Foster is also skeptical. He rejects, "the dream that technology alone, considered in some abstract sense, can solve the environmental problem, allowing for unending economic growth without undue ecological effects through an absolute decoupling of one from the other" (Foster and Clark 2012). In a similar vein, Moore worries about modernist ideas of humans controlling nature and the renewed Prometheanism of the Anthropocene argument wherein the old formula IPAT (Impact = Population*Affluence*Technology) returns, simply with invigorated concern about the level of Impact (Moore 2016).

Foster, especially with Clark (Clark and Foster 2012), similarly shows how the increasing scales of technology of agriculture led Western producers to need new fertilizer inputs and, with scientific technological developments, to turn to guano imports from South America. The result shored up Northern agriculture for some decades, while degrading the landscape of Peru and exploiting Chinese labor in the process. Moore, especially in his historical examinations of the world-system and its expansionary tendencies, such as the "progress" of sugar production from Madeira to the Caribbean and Brazil, also shows how particular commodity frontiers can expand with technological developments and labor inputs but are limited by "physical geography and the contradictions of capitalism" (Moore 2000:412).

Third, they agree on the causal inability of capitalism to offer solutions to our current societal and global dilemmas. To the contrary, in fact, they agree about the fundamental role of the capitalist world-system in creating the unsustainable and exploitative practices that characterize our world. Bunker beseeched us in the 1980s to abandon the internal–external debate on underdevelopment and articulation of modes of production to understand the regional dynamics of extractive peripheries as they are incorporated into wider circuits of trade, facilitated by collaborative states in the periphery. For Foster, the development of ecological imperialism into a global understanding of the metabolic rift is founded in capitalist exploitation. Moore (2014) labels the intertwined relations of capital and nature over the last 500 years, the age of capital because capitalism has mostly not paid for the real costs of the endless accumulation of capital.

Finally, and most crucially for the discussion in this volume, in my view they fundamentally agree about the relations of inequality embodied in the concept of EUE. Bunker's contribution was foundational in relating ongoing debates about economic unequal exchange in critical development studies to include the *ecological*. Rejecting a wage differential between core and peripheral areas as the sole cause of unequal exchange (per Amin), he wrote, "The appropriation of values in nature, from the periphery, in fact initiated unequal exchange between regions, and between ecosystems, long before the rise of wages and the expansion of consumer demand in the core" (Bunker 1985:45; see Bunker, page 29, this volume). Foster, too, sees EUE operating, although he focuses on "Marx's use of the concept of metabolism to account for the human-nature relation through social production" (Foster 2000:159). In *Capital*, Foster finds Marx's insight into how capitalism robs both workers and the soil. He notes that for Marx, "the disruption of the soil cycle in industrialized capitalist agriculture constituted nothing less than 'a rift' in the metabolic relation between human beings and nature" (Foster 2013). This rift, furthermore, is global and the "worst forms of degradation, as well as the pillaging of resources and the disruption of sustainable relations to the earth, are concentrated in the periphery" (Clark and Foster 2012:69). Clark and Foster thus conclude, "Unequal ecological exchange, an outcome of the global metabolic rift, has allowed for the growth of the center of the system at unsustainable rates." While Moore (2011b:112, original emphasis) only cautiously allows that Bunker (and others), "shed light on the ways that biophysical transformations have *enabled* accumulation, and capitalist development as a whole," his work on the sugar frontiers is in fact quite compatible with EUE. "On balance," beginning in the fifteenth century, Moore writes, "urban-based capital looked abroad for new landscapes where the 'original sources of all wealth' (land and labor) could be mobilized in servitude to the commodity form" (Moore 2010a:3). The expansion of sugar production on Madeira island represented a "monocultural strategy [that] was progressively self-limiting" due to weed and pest problems (Moore 2010a:8–9). As a result, capital left Madeira—and the Amazon, as well as Peru's guano islands—behind as "regional crises were resolved through global expansion" (Moore 2010a:19).

In delineating "multiple inequalities in international exchange," Bunker's research emphasized the spatial inequalities between industrialized modes of production and peripheral modes of extraction. Combining the conditions of production concept of O'Connor (1988) with the socionatural concept of Swyngedouw (1999), one might say that these are differences in the "socionatural conditions" of more modern and industrialized modes of production in the core of the world-system and less modern peripheral modes of extraction. Due to the extraction of "natural value in the raw resources" in extractive peripheries spatially and socially distant from "the location of the full realization of value," peripheries such as the Amazon suffer from repetitive socionatural impoverishments (Bunker 1985:45; see Bunker, page 29, this volume). Yet, importantly, Bunker beseeched researchers to avoid "reifying dependency, unequal exchange, or capitalism as causal agents" (Bunker 1985:50; see Bunker, page 34, this volume). Clark and Foster (2012) also examine the spatiality of soils being "robbed" and coming to rely on guano from islands off Peru's coastal extractive periphery. This case and others are used by Foster (2013) to argue that metabolic rift theory is "a theory of ecological crisis—of the disruption of what Marx saw as the everlasting dependence of human society on the conditions of organic existence." Along "commodity frontiers," Moore finds there was a "capitalization of socio-ecological relations ... joined to the appropriation of ... a very large basket of nature's gifts." Madeira experienced "crisis" (in Moore's history due to labor productivity declining) and became "but one leg of a great frontier journey" of Portuguese imperialist expansion just as the Amazon was too during cycles of commodification from turtle oil to rubber and beyond. To be sure, one difference is that Bunker retains his focus on the zone of extraction within world-systemic processes, and the Amazon and later the Brazilian state retain agency in the process. Meanwhile, Moore (2010a:4, 2015a) follows commodities like sugar to new frontiers and frames his discussion in terms of the more expansive "conditions of reproduction (the web of life)." As with Foster, then, the focus is on the dynamics of the capitalist system almost more than the ecological, demographic, organizational, and infrastructural dynamics in particular peripheries and the importance Bunker found in the national state.

Beyond these foundational affinities in seeing extractive regions as experiencing EUE via unequal power relations, extraction of "free gifts" of nature to support the expansion of capitalist accumulation elsewhere, and the deep and persistent underdevelopment of the extractive peripheries, there are some distinctions that represent real differences among their approaches. Their ideas differ, first, about *how capitalism causes degradation* of the human and nonhuman parts of nature. Fundamentally, second, this difference relates to theoretical and especially ontological differences of perspective about *what "nature" is* and whether it is separate from society. Third, the ontological differences are reflected in epistemological differences about *how we know* nature, including debates about dialectical method. Fourth, these differences all have implications for the *kinds of futures* envisioned and ideas about how to reach such futures. Before reviewing these intellectual differences, I return briefly to the unfortunate sectarianism on the environmental left that has clouded potential debate.

Unfortunate Sectarianism

Debate, ideally, is supposed to bring precision or even enlightenment, but in some circumstances it does not. Sometimes, questions of power intervene. Other times there is insufficient deliberation. Sadly, in my view, intellectuals who should be allies on the green-red axis have devolved into throwing darts,[6] except one side—represented by Jason Moore, along with his colleagues and students in the world-ecology group (e.g., Ben Marley, Christopher Cox, Raj Patel, et al.)—seems to be the recipient of more numerous and personal attacks than the other side—represented by John Bellamy Foster and his colleagues and students in the metabolic rift group (e.g., Brett Clark, Hannah Holleman, and others). As a result, there is tremendous unrealized potential for a fruitful and useful debate about society, nature, metabolic rift, world-ecology, and so on, and their important relation to EUE to capture the unequal ways in which different parts of the world are affected by the (inter)relations of humans and the rest of nature.

After several years of not responding at all to Moore's numerous publications and evolving world-ecology approach, Foster did so first in some *Monthly Review* columns and an interview on the blog "climate and capitalism." In the interview (Angus 2016), Foster described Moore in the most strident terms: as "opposed to ecosocialism and to the radical ecological movement in general," having an outlook "wrapped up" with the Breakthrough Institute in denying environmental degradation and climate change, "only ironically" Marxist, and, in the end, having "moved to the other side." Although Moore has been relentless in his criticisms of other perspectives, I have never read him accusing Foster and others of such things. What Moore has criticized the most consistently has been the Cartesian dualisms that he sees riddled through left environmental critique. He has been perhaps too cute in his language of summarizing these problems as "Green Arithmetic," but his points have been ontological and epistemological.

Foster has been understandably bothered by the emphasis on language in Moore's rejection of capitalism's "war on the environment," but, at the same time, language does shape understanding. The repeated usage of a discourse of ecological impacts and footprints gives the sense of a putative separation of humans from nature. Moore's response is to offer "monism" as part of a "world-ecology" perspective that is presented as open to wider discussions with multiple disciplines and more open-ended about possible futures, as I show below.

Yet, there are still moments when both seem open to further debate. For example, Foster and Holleman (2014:200) declare, "Our goal here is merely to open the door to what we hope will eventually be a comprehensive theory—one that would need to be integrated with issues of history, geography and co-evolutionary development, encompassing the whole formation of the world-capitalist system, including its historical logic and crises." Similarly, Moore prefaces his *Web of Life* by writing, "This book is an invitation. It is offered as an opening to conversation, and an incitement to serious debate, over humanity's place in nature, and how our thinking about this place in nature shapes our view of history, our analysis of the present crisis, and the politics of liberation for *all* life" (Moore 2015a:ix, original emphasis).

In what follows, I take this opening seriously and try to stick to the real distinctions between their positions. I do so by juxtaposing them to Bunker, who not only offered foundational ideas on EUE but had useful contributions to the distinctions being made now.

Distinction I: How Capitalism Causes Degradation

In order to understand the views of Bunker, Foster, and Moore on the relationship between capitalist expansion and (socio)natural degradation, it is worth recalling their respective objects of inquiry. In other words, what is the explanandum of each of them? In brief, it is regional development, the global metabolic rift, and historical nature, respectively. These distinct foci shape each scholar's understanding of how capitalism causes degradation.

In *Underdeveloping the Amazon*, Bunker declared his interest in understanding regional development, especially "uneven development between regions and the capacity of one region to subordinate another" (Bunker 1985:245). To be sure, it is the arrival of external merchants that creates the earliest historical decimation of human and nonhuman populations. Importantly, he stressed the evidence that indigenous populations of the Amazon region before Europeans arrived were larger and more stable than subsequent Western histories admitted. Failed efforts to establish sugar plantations further inland, including aggressive slave raiding campaigns and the spread of disease, reduced the native populations, but it was the extraction of turtle oil and manatees that decimated them by reducing their sources of oil and meat while "disrupting critical links in the riverine ecosystem" (Bunker 1985:64; see Bunker, page 41, this volume). Moreover, "Europeans' rapacity, and their stubborn belief that as members of a master race they should not engage in productive work rapidly exhausted the resources on which their dreams of great wealth were founded." The twin results were local poverty and resource exhaustion that were worsened by the subsequent cycle of rubber extraction. While many, including Moore, remember Bunker for stressing extraction of value in nature, Bunker recounts how rubber boomed on the backs of imported tapper labor and collapsed amid Malayan competition, due to both disease and lack of sufficient rural population to establish plantations in the Amazon. This situation set the stage for the twentieth century minerals extraction that further degraded and subordinated the Amazonian region to Brazilian state and corporate elites who in turn are dependent on global markets for the extractive revenue-generating activities.

Considering the historical cycles of extraction, Bunker's approach to understanding capitalism's relation to regional development was complex, field-based, and multi-scalar. It was also theoretically eclectic. As he reached the conclusion to *Underdeveloping the Amazon*, he reviewed his selective borrowing from theories of modernization and of modes of production "to explain the consequences of certain uses of both the natural and the social environment within a particular region" and from world-systems and dependency, especially at the global level, to account for the "world market as systemic and comprising multiple regional economies, world system as dominated by core economies, and world system as accelerating accumulation in a small part of the world." In addition, the power or capacity of one region to subordinate another was attributed to "energy flow-through in social organization and infrastructure" which Bunker views as limited in extractive peripheries such as Brazil's Amazon (Bunker 1985:244–245).[7]

Foster's intellectual project is different in that he aims to understand the metabolic rift.[8] The metabolic or ecological rift, Foster and colleagues argue is, "the product of a social rift: the domination of human being by human being." At the global level, it is an imperialist project in which, "This larger world of unequal exchange is as much a part of capitalism as the search for profits and accumulation" (Foster et al. 2010:47). They sum this relationship up by rephrasing O'Connor's second contradiction of capitalism as the "absolute general law of environmental degradation under capitalism" (Foster et al. 2010:207) which involves wealth accumulation at one pole and "the accumulation of conditions of resource depletion" at the other. Putting this in the context of the growth of monopoly capitalism and crisis since the 1970s, capital is viewed as responding by "a speedup of the destruction of the remaining natural forest ecosystems throughout the world" (210). Speaking self-consciously for socialist ecologists, Foster's diagnosis is that capitalism, "has generated an acceleration of the human transformation of the Earth system in two major phases: (1) the industrial revolution beginning at the end of the eighteenth century and (2) the rise of monopoly capitalism, particularly in its mature stage following the Second World War" (Foster 2015:7). He usually does not write from a particular geographical location as Bunker did but instead builds a more general analysis from multiple locations and multiple ecological crises or tipping points (see Foster 2009, 2013). In building this theoretical understanding, Foster emphasizes that those capitalist and perhaps core workers who are deluded into believing they benefit from the current situation are sorely mistaken because they too suffer from the rift.

Moore is distinct from both. He aspires to understand historical nature. In this effort, he expends most of his effort analyzing what capitalism actually does to the world, what he dubs the web of life (Moore 2015a). From this effort, he finds, "Capitalism's governing conceit is that it may do with Nature as it pleases, that Nature is external and may be coded, quantified, and rationalized to serve economic growth, social development, or some other higher good. This is capitalism *as a project.* The reality-the *historical process*-is radically different" (Moore 2017a:601, original emphasis). In a footnote that could apply to Bunker, Moore notes that those focused on regional (as opposed to global or systemic) change do not mistakenly operationalize capitalism as operating as it pleases on an external nature. Also, Moore emphasizes how capitalism produces "cheap" labor, food, natural resources, and energy, while Bunker and Ciccantell explain *how* raw materials are cheapened through increasing economies of scale and the creation of "generative sectors" (2005:84–86).

In taking up the lessons of his world-ecology analysis of sugar frontiers, for example, Moore rejects the characterization of soil fertility as an "external" barrier. Rather, "soil fertility and exhaustion are in fact eminently historical relations internal to the capitalist mode of production"

(Moore 2010a:6). While understandable for soil and agriculture, this position becomes a bit more complex in the case of minerals and other more purely extractive parts of commodified nature. Yet even in such cases Moore insists that the combinations or "bundles" of human and nonhuman parts of nature vary widely and need to be analyzed historically. Bunker seems to agree but adds specificity in his assessment of mineral extraction and human technologies. As he observed, "The laws of physics and chemistry and the laws of motion of capital both constrain the development of technology" (Bunker 2005:39). Moore's way of handling the shaping of technologies is to examine co-productions, writing "Capitalism takes shape through the co-production of nature, the pursuit of power, and the accumulation of capital." (Moore 2015a:46).

While their understandings of the analytical separability of human activity from (the rest of) "nature" differ, the pursuit of power and capital accumulation brings Bunker and Moore together. Harkening back to Arrighi and Braudel, Moore writes, "Agency, limits and crises—*but also 'golden ages'*—are co-produced by human organizations" (Moore 2017b:601, original emphasis). In other words, capitalism is a system that creates winners—core nation states with opulence, consumption, arts, and creativity—and losers—extractive peripheries with poverty, precarity, exhaustion and limited opportunities for future dynamism. This conjoining of the social and the spatial in an understanding of capitalist exploitation of labor and nature (together) may explain why it is so hard to convince even subordinate class members in the dominant regions of the world that they are compromised by a metabolic rift. It is a point that Foster would do well to include more clearly in his analysis.

Distinction II: On Nature's Ontology

The second distinction is regarding the ontological position of nature in one's analysis. To wit: Bunker saw value in nature, Foster clarifies this value in nature as wealth based on use values, whereas Moore views nature as a tangled matrix. While Bunker's theorizing makes nature seem to have the most separate ontology, his field-based investigations led him to depict the intertwined complexities more fully than the others.

To begin, Bunker insisted on value resting in nature. “When natural resources are extracted from one regional ecosystem to be transformed and consumed in another, the resource-exporting region loses values that occur in its physical environment,” wrote Bunker (1985:22; see Bunker, page 15, this volume). Bunker and Ciccantell (2005:3) put it even more simply, “Humans cannot create matter, so they must extract it from natural sources.” In part their perspective is understandable due to their focus on mineral raw materials where the question of ongoing or “sustainable” production is not relevant whereas Moore and Foster both stress cases of agricultural expansion. For Bunker, natural value has importance in the ways it is integrated into an unequal world-system. Those regions that dominate the world are compelled within a capitalist system “to assure the increasing, stable, and cheap supplies of the raw materials” that support them (Bunker 2005:41) This domination is accomplished in extractive peripheries via a process of appropriation that relies on power and violence. In the Amazon, Bunker found, “appropriation impoverishes the environment on which local populations depend both for their own reproduction and for the extraction of commodities for export” (Bunker 1985:22; see Bunker, page 16, this volume).

For Foster, too, nature is real and worthy of understanding on its own. “Appropriately,” Foster et al. (2010:250) write that social scientists are increasingly concerned with “the intersection of human society and nature.” Their worry is that “Little time is spent understanding natural processes and patterns: how they operate on their own, how historical social systems interact with nature, how nature influences social conditions, and how natural processes are transformed by social interactions” (251). In sum, Foster (2017:2) finds that “the metabolism of nature and society” *is* the organization of production.

In terms of value, as those familiar with his long career in the *Monthly Review* or monopoly capitalism school of thought know, Foster embarks from a classical Marxist position based in the labor theory of value. As a result, he is critical of Bunker for his “inconsistent allusions” to Marx and “an undefined theory of ‘natural value’” (Foster and Holleman 2014:211). The former point can be rejected out of hand; it appears to be Foster’s way of simply saying that Bunker (like Moore) is not a good enough or pure enough Marxist. It is the latter point that is more important. Can nature have value and if so what kind? In *Marx’s Ecology*, Foster observes that Marx did not reject the idea of nature having value but linked it to his distinction between use value or "genuine wealth" and exchange value. "Nature... was just as much a source of wealth as labor" (Foster 2000:168). It is capitalism, in this view, that causes nature to have no value. Foster et al. (2010:63) summarize the situation as follows: "Those who—falling prey to the commodity fetishism of capitalist value analysis—saw labor as the sole source of wealth were thus attributing 'supernatural creative power' to it."

To capture this real wealth as opposed to capitalist (exchange) value, Foster has taken up Odum's idea of "emergy" or embodied energy in a recent article with Holleman. They praise Odum for his consistent application of emergy:

> 'Emergy', he stressed, 'measures natural value—real wealth' (Odum 2001:112). Not only was money not a measure of real wealth, the relation was often an 'inverse' one, with prices 'being lowest when [ecological] contributions are greatest' (Odum 1991:90). The whole analysis pointed to a notion of 'emvalue in a value added hierarchy' that resembled Marx's analysis but was oriented instead to real wealth—seen as in contradiction with the labor-value (or human- services) basis of the capitalist economy. (Foster and Holleman 2014:214)

However, as much as one might try, they note that emergy cannot succumb to a single metric. Perhaps that is why Bunker gave up on the effort (begun in *Underdeveloping the Amazon*) to calculate EUE in simple energy terms.

In his anti-dualist stance, Moore agrees with Foster on the inseparability of exploiting (or appropriating) nature and exploiting labor in the capitalist mode of production. Schneider and McMichael (2010:479) explain, "Moore's comment, in distinguishing capitalism's value form from nature's wealth—'Marx does not deny that external nature does work useful to humans, only that (*from the perspective of capital*) its productions do not directly enter into capitalism's particular crystallization of wealth' (2003:450, original emphasis)—exemplifies the determination of wealth under capitalism by value relations as an ontology." Moore tries to push the point further to a relational ontology. What he calls a

Cartesian dualism of society (or capitalism) and nature, "confuses particular natures that are objects of capitalist development with nature as the matrix within which capitalism develops" (Moore 2015a:21). Moore's solution is to insist on a monist ontology of this matrix that cannot be untangled. In this monist view, nature is never separate. Thus he writes:

> the difference [is] between "capitalism and nature" and capitalism-in-nature, whereby the accumulation of capital and the production of nature become so intertwined that the one is unthinkable without the other. "Nature" is no longer a passive substance upon which humanity leaves its footprint. Rather, it becomes an inclusive and active bundle of relations formed and re-formed through the historically- and geographically-specific movements of humans with the rest of nature. (Moore 2011b:119)

This forming and re-forming is not completely new. It is something that Bunker also saw in his historical analysis of complex ecologies among the turtles, manatees, and humans. As such, human intervention, "does not constitute the control supposed in the conventional historical myth, nor does his failure to control the environment end in the return of nature to its primeval state. Rather, each human intervention in the environment transforms it in ways which limit the possibilities of subsequent interventions" (Bunker 1985:14).

In addition, Moore (2017a, 2017b) adds further complexity to his rejection of dualism and his analysis of exploitation of an inseparable capitalism-in-nature with the idea of appropriation. Specifically, he writes, "Value operates through a dialectic of exploitation and appropriation that illuminates capitalism's peculiar relation with, and within, nature" (Moore 2015a:16). In addition to the commodification of labor that is fundamental to exploitation, Moore adds appropriation, which he says, "works—and this is less widely understood—by elaborating forms of power, re/production and rationality that mobilize work in service to capital, but outside profit/loss accounting (accumulation by appropriation)" (Moore 2017b:606). This point is strikingly similar to Bunker, who had written that the differences between productive and extractive regional economies, "create unequal exchange not only in terms of the labor value incorporated into products but also through the direct appropriation of rapidly depleted or nonrenewable natural resources" (Bunker 1985:22; see Bunker, page 16, this volume). One might plausibly make the point that Bunker's account of the Amazon did not take into account feminist theories of reproduction as Moore increasingly tries to do, but he did recognize appropriation, including the demographic impact of early Western contact from violence and disease on the extant human populations of the Amazon and the Andes (Bunker 1985, 2005). Where Moore remains distinct is in taking a relational, capitalism-in-nature view of even geological matter like coal to argue that while theorists have accepted the ontological inseparability nature, "historiography of resource extraction has seldom taken the relational point seriously" (Moore 2015a:196, 2017c:19). Yet, if something (nature?) can be appropriated, one wonders why Moore does not think that there is any analytical moment to consider the appropriation and the relationality separately. We turn to the issue of knowledge next.

Distinction III: Dialectics or How to Know Nature

We have seen that Bunker and Foster retain a firm ontological separateness of nature, or for Bunker simply matter, while Moore argues for inseparability. Others such as White, Rudy and Gareau, have noted, "Metabolic rift theory emphasizes the dialectical nature of socio-ecological relations but ... leans to a view of 'nature' as *external from capitalism, singular and largely fixed*" (White et al. 2015:104, original emphasis). Schneider and McMichael (2010), too, have questioned the dualism of the metabolic rift and argued for reunification of the social and the ecological. Furthermore, Schneider and McMichael make the point that the encounter between different systems such as agroecosystems, and similarly, one might add, Bunker's productive and extractive economies, is messy because each "side" is not static and fully organized. Yet, "scholars avoid these empirical problems in part by discussing the metabolic rift in the abstract, using the language of 'nutrients' and 'nutrient cycles'" (Schneider and McMichael 2010:474). These observations fit the ontological weakness of Foster's position and the strengths of Bunker's empirical analysis. That is why Moore's ontological position is exciting and enticing; it captures the dialectical relations between humans with(in) nature.

However, the next question is, how can we study, analyze, and know nature? This epistemological question leads us into the intricacies of dialectical analysis. It is interesting to note the difficulty of analyzing (or writing) about "nature" in a nondualist or monist manner. Even Moore, despite strenuous efforts, occasionally writes in a dualist modality in passages such as this one: "Deforestation weighed particularly heavily on highly vulnerable mountain eco-systems, which suffer from high rates of soil erosion …" (Moore 2010b:48). On the epistemological distinction, I am more convinced by Foster's view of dialectics than Moore's. Foster is more in line with the oeuvre of Bunker, too, who recognized the need for abstraction and temporary isolation of elements.

Foster (2013) explains that there is "nothing dualistic or non-reflexive" in his social metabolism position from an epistemological perspective. Instead he posits that Moore and other "left critics" have misunderstood what dialectical thinking is truly about. Foster (2016:404) observes that in Moore's quest to dash any dualism, he "seems to think that one cannot speak abstractly, as Marx did, of a metabolic relation of humanity to the earth through production, i.e. a social metabolism, while also recognizing the universal metabolism of nature within which this social metabolism necessarily exists." Analyzing the anti-dualist monism of Moore and others, Foster writes, "What is most often missing in … social monism is the understanding of complex mediations between nature and society within a dialectical concept of totality" (Foster 2016:399). He concludes that monism, "subsumes the environment within society—in effect abandoning the dialectic of nature and society by reducing the former to the latter" (401).

More to the point, Foster identifies, "a failure to perceive that within a materialist-dialectical perspective it is impossible to analyze the world in a meaningful way except through the use of abstraction which *temporarily isolates*, for purposes of analysis, one 'moment' (or mediation) within a totality" (Foster 2013, emphasis added). In analogous manner, Bunker (1984:1021, emphasis added; see also Bunker 1985:49), while focusing on separate spaces and modes of production, also found, "A full account of the intersection between internal and external systems requires *separate analysis* of each system in terms which recognize the dynamics of each system as an integral unit and simultaneously permit analysis of their effects on each other." These temporary and separate methodological moves require re-integration at the moment of theorizing, to be sure.

Foster's incisive mode of Marxist dialectics insists on the temporary isolation of "natural" and "social" systems as moments within a broader totality. Moore aims in his historical approach to reject this view as re-creating Cartesian dualisms. He believes this "emphasis on the interaction of positions confuses the results of the dialectic of human and extra-human natures with the dialectic itself. This dialectic, what I call the *oikeios*, gives rise to the singular abstractions of 'nature' and 'society' in their specifically capitalist forms" (Moore 2011a:2). Thus, for Moore, it seems that dialectics refers to the ontological unity of the *oikeios* or the "web of life" (Moore 2015a, 2015b), which is torn asunder by capitalism. To Foster that is precisely what the metabolic rift is, i.e., the tearing apart of previously unified parts of the web of life, for which he prefers Marx's term the "universal metabolism of nature." However, when Moore argues that the capitalist world-economy acts *through* an irreducibly socialized nature, it is hard not to see the idealism that Foster identifies in Moore's recent work. Moore (2016:2) claims, "modes of thought are tenacious. They are no easier to transcend than the 'modes of production' they reflect and help to shape." Furthermore, Moore rejects the accusation of idealism, deflecting it to capitalism itself as a "project" whose "imagination is vigorously constructivist" while stressing that "one must 'see like capital' in order to transcend the illusions of capital" (Moore 2017b:601).

Yet, by drawing attention to the impossibility of separation—even temporarily for purposes of analysis—Moore's monist approach poses insurmountable barriers to sociological research. But this distinction, to reiterate, is an ontological one, not an epistemological and dialectical one. As Foster explains, the issue is one of dialectical analysis and one's method of abstraction. In Moore's quest to dash any dualism, he "seems to think that one cannot speak abstractly, as Marx did, of a metabolic relation of humanity to the earth through production, i.e. a social metabolism, while also recognizing the universal metabolism of nature within which this social metabolism necessarily exists" (Foster 2016:404). Of necessity, one must employ a dialectical method by:

> employing conceptions that at first sight—when separated out from the overall dynamics—may appear one-sided, mechanical, dualistic, or reductionist. In referring, as Marx does, to "the metabolic interaction between nature and man" it should never be supposed that "man" (humanity) actually exists completely independently of or outside of "nature"—or even that nature today exists completely independent of (or unaffected by) humanity (Foster 2013).

After temporarily separating out "society" and "nature" as analytical abstractions, the key is that "we have to join them again, show their interpenetration, their mutual determination, their entwined evolution, and yet also their distinctness" (Holleman 2015:6). Re-reading Bunker's analysis of the Amazon, it strikes one how detailed his analysis of mutual determination was (Gellert 2005). Perhaps Foster and his colleagues might convince Moore if they made more effort to do that latter joining effort.

One final point on dialectics is that it includes not only analytical movement between abstract and concrete but also reflexive analysis from particular vantage points (Ollman 2003). Many have tried to re-join the disjointed, moving beyond society and nature to "think like a mountain," but few have truly done so (Freudenburg, Frickel, and Gramling 1995; Gellert 2005; Goldman and Schurman 2000). One of Bunker's contributions to this epistemological dilemma was to take multiple vantage points in his work. Bunker and Ciccantell (2005) appear to take the global or bird's-eye view of capitalist cycles of accumulation but still integrate local material processes. Earlier, in *Underdeveloping the Amazon*, Bunker had made it clear that local experiences and practices are vital and that the local is not a separate location from the global. One might say that like Moore's insistence on capitalism-in-nature and nature-in-capitalism, Bunker insisted on global-in-local and local-in-global. Moore, too, recognizes the importance of vantage point, for example, in distinguishing different periodizations of the origins of capitalism, asking: "Are not these different interpretations premised on distinctive angles of vision?" (Moore 2017a:621). In his posthumously published final work on Andean irrigation before Europeans arrived, Bunker (2006) revealed that he uncovered the "snake with the golden braid" in the Andes Mountains by spending time walking and feeling the mountains while thinking like and talking with some of the indigenous people. The snake, he believed, was simultaneously "the animal manifestation of the ditches, of the water in them, and then increasingly of the understandings of landscape and water flow that had been bestowed on the builders of the original ditches" (Bunker 2006:xii). He concluded (2006:112) that, "continuities from the Amazon to the Andes in the relations of society to nature strongly suggest the futility of attempts to describe these relations in linear, dichotomous, or unidirectional terms. Arguing about whether the environment shapes society or whether society constructs nature misses the essential lesson that society and nature interact from very different material bases, organizational forms, reproductive dynamics, conceptual frames and temporal systems, but that they do so in reciprocally formative ways."

Distinction IV: Crises and Future Visions

Envisioning the future is always a challenge for social theorists. Arguably the deepest division among the three thinkers addressed in this chapter is their view of the future. Bunker's work was framed in the dependency-influenced idiom of the development of underdevelopment, but he insisted on the importance of overcoming the divide between so-called external and internal causes of underdevelopment, and his vision calls for more local or autocentric development with humans working more in harmony with other parts of nature. Foster's work is centered in Marxist analysis and an ecosocialist critique that calls broadly for a particular kind of social revolution. Moore's work sits within critical environmental history and humanities and envisions multiple future ways of constructing a new *oikeois*. In this section, I examine the ways each of them evaluates the current trajectory of socionatural affairs and their vision for the future. In order to understand the future, one has to briefly consider their views on the ecological crisis (or crises) and who or what is facing it (or them) so I begin there.

Bunker's concern in *Underdeveloping the Amazon* was with the rural people and places in that region. Due to over 350 years of extractive exploitation and EUE, the Amazon and its peoples face "ecological devastation" (Bunker 1985:250). "The dilemma," he concluded, "is that self-sustaining, symbiotic economies adapted to and exchanging across different ecological zones can only be maintained in a system that permits ecosystem maintenance and long-term conservation to prevail over short-term profit maximization" (251). That is why he highlighted the multiple failures of the Brazilian state and rejected the possibility of its interventions leading to anything but further environmental and social disruption. Moreover, in recognition of the inseparability of humans from the rest of nature, he appreciated how this process impoverishes dominant elites and "the nation as a whole" (249). In Bunker and Ciccantell's (2005:241) examination of the race for resources they concluded that there is power in "unveiling" the vast inequalities produced via EUE and thereby convincing peripheral states to combat rather than facilitate extraction by powerful core states and corporations. Yet in his earlier book he had noted that because (modernizing) states are limited to "only repressive and allocative powers," one needs the "complete transformation of class relations in a socialist revolution" and "even socialist states are affected by prior social and economic formations" (Bunker 1985:241).

Bunker found the self-sustaining economies he was looking for in the Andean mountains and described them in his little-noticed last work *The Snake with the Golden Braid* (2006). Finding that complex Huanoquite community irrigation systems could and were locally organized and democratic—rather than inherently supportive of despotic and authoritarian rule, as Wittfogel had claimed, led him to an alternative interpretation based on "close attention to the materio-spatial characteristics and social organization requirements of mountain irrigation systems" (Bunker 2006:110). The Incan system of the Huanoquitenos based in locally specific knowledge and authority was far superior to the Spanish absolute rule that destroyed it. By uncovering their cosmology and their material practices, he saw how the people "use, maintain, abuse, forget, or neglect the technologies" and argued that the Huanoquite system "can teach us a great deal about the evolving interactions between society and nature, and perhaps about ways we humans can manage the social side of that interaction better to reduce its damage to the natural side" (Bunker 2006:2). Finally, he believed that the kind of future that he desired—for all of us—was closer to what he found there, one that was "far more open to locally specific authority and knowledge [and was] knowledge-based, not power-imposed" (Bunker 2006:110).

Foster's understanding of crisis is more generalized, as is his vision for the future. There is another distinction in this case between Foster and Moore because Foster accepts the science on the ecological crisis that we are facing while Moore worries about the apocalypticism of Foster's view. In accessible writing, Foster and Magdoff's (2011) "citizen's guide" to *What Every Environmentalist Needs to Know About Capitalism* delineates environmental indicators that could easily have been plucked from World Watch's *State of the World* series or Sen. Al Gore's *An Inconvenient Truth*. In this view, melting ice caps, rising sea levels and temperatures, and warming global temperatures are indicators of *the* environmental crisis reinterpreted as "planetary rifts."[9] In this telling, capitalism (or more benignly, simply "human production") faces resource limits and waste limits with the result that, "All ecosystems on Earth are in visible decline" (Foster and Magdoff 2011:77). As Foster and Clark (2012) delineated them, "Ocean acidification, destruction of the ozone layer, species extinction, the disruption of the nitrogen and phosphorus cycles, growing fresh water shortages, land-cover change, and chemical pollution all represent global ecological transformations/crises."

In facing the ultimate crisis of human survival, Foster insists that there is a vast consensus on the ecological left on the importance of Marx's insights into the metabolic rift. As he summarizes the current situation, "The rediscovery of the ecological value-form character of Marx's political economy, his conception of metabolic rift, and his recognition of unequal ecological exchange (and ecological imperialism) have all shifted the ecological debate globally in more revolutionary directions" (Foster 2016:397). Foster (2017:11) advocates that we build an "ecological civilization in the Marxian sense" which would "transcend the logic of all previous class-based civilizations, and particularly capitalism, namely, the interconnections between the domination/alienation of nature and the domination/alienation of humanity."

Yet, the universalist and apocalyptic understanding of our current state of affairs leads Foster (and *Monthly Review* authors more generally) to call—almost out of reflex—for a socialist revolution that will bring "the rational, social regulation of the human metabolism with nature envisioned by Marx" (Foster and Clark 2012). Because there is "no possible way to accomplish any, much less all, of these things other than by breaking with the underlying logic of the accumulation of capital" we need "a very rapid overthrow of capitalism of a kind that can scarcely be imagined today" (Foster and Clark 2012). Foster's point is that capitalism is creating a widening "rift" in the metabolism governing our lives. His conclusion is that the metabolic rift and now planetary rift of capitalism "can only be surmounted by socialism, based on a rational, sustainable, relation to nature" (Angus 2016). However, there is scant mention of how to get there from here. Moreover the pleas for rational organization of life lead critics like Moore (2016) to recoil back from the renewed reliance on a kind of "planetary engineering."

In Moore's rejection of such apocalypticism there is not exactly the naivete that Foster thinks. Rather, more like Bunker, he stresses the need to think carefully about *historical* natures (Moore 2015a:19). It is this attention to place, space and time that distinguishes Moore's (2016) rejection of the anthropocene and his vision for the future.[10] Moore's building on the shoulders of metabolic rift giants is found in this passage, in which he recognizes the same "natural" processes that Foster dwells on but comes to a somewhat distinct conclusion:

> And if the limits of capitalism today are limits of a particular way of organizing nature—this is hardly to deny the acceleration of biospheric change through global warming, the Sixth Great Extinction, and more—then we are confronted with the possibility of changing humanity's relation to nature, which is to say also humanity's relation to itself. (Moore 2014:17)

Somewhat surprisingly, however, Moore places his bets for the future not on social revolution or ecological revolution but on capitalism's decreasing ability to find cheap commodity frontiers after five centuries of incredible creativity in that regard. "There are limits," he concludes, "to how much new work capitalism can squeeze out of new working classes, forests, aquifers, oilfields, coal seams, and everything else. Nature is finite. Capital is premised on the infinite" (Moore 2014:17). Like Foster, he does not offer a full politics of praxis that might get us from here to there. Moore's resolution is decidedly idealist, declaring, "Efforts to transcend capitalism in any egalitarian and broadly sustainable fashion will be stymied so long as the political imagination is captive to capitalism's either/or organization of reality" (Moore 2015a:2).

Conclusion

In this chapter, I have tried to spread more light than heat on the fundamental similarities and important distinctions among Bunker, Foster, and Moore. Bunker's work on EUE remains foundational for good reason, I argue. In their works, both Foster and Moore basically assent to the insight that more powerful regions dominate other regions, especially extractive peripheries, through the appropriation of nature and the exploitation of labor to do the appropriating. They build on this foundation to pursue their research projects in distinct directions, however.

The main distinctions are as follows. First, in terms of capitalism's role in degradation, despite the broad agreement on EUE, Bunker's peripheral vantage point grounds his perspective in the soils of appropriation and exploitation the most fully. Moore, in my view, continues the most directly in this direction, notably in his analysis of sugar frontiers, in developing a world-ecological view of the last 500 years. Foster, by contrast, remains wedded to the idea that human impacts on "nature" so greatly intensified during the industrial revolution that one must examine the metabolic rift in this more recent period. Second, in terms of ontology, Moore rejects this view of human impacts and pushes the furthest toward a fully socionatural ontology that refuses to disentangle the web of life into constituent parts. In his monist view, there is no real ontological separation of humans from other parts of nature. Bunker had hinted at a similar ontology both in his refusal to consider the remotest reaches of the Amazon as separate from the centers of capitalist hegemony and in his self-reflective analysis of the cognitive and physical impact of terrain and irrigation on his analysis and importantly on the capacity of pre-Conquest indigenous Andean people to conceive of and create complex irrigation systems. Third, while ontological monism breeds insights, it may falter methodologically. Here Foster's epistemology is more useful. His consistent application of a dialectical method to temporarily disentangle elements in the complexity or web of life that have been separated by the metabolic rift avoids the insurmountable problem of a researcher deciding which elements of socionature are entangled and how at particular moments in time. In this regard, Bunker's analyses, such as of the surprising population dynamics of turtles, manatees and humans, had provided an early exemplar of how to not just temporarily separate but importantly re-combine the elements.

The final area of distinction among these authors relates to options and inflections of how to pursue a future that is (for want of a nontrite expression) more sustainable. Here the academic terrain, I argue, is rather flawed. Scholars worry about whether they are being "optimistic" or "pessimistic."[11] We need not. To mix metaphors from Weber and Marx, our vocation is to analyze the world and seek ways to change it. In this challenge, Foster's vision is expressed in the most theoretically restrictive way (or precise, depending on your point of view). Moore's vision is more expansive but also more idealist. Bunker's vision through most of his work, including his foundational work on EUE, is realistic from the vantage point of the extreme periphery. Yet, his last work offers a glimpse of a more embodied understanding of the ontologically inseparable socionatures toward which we can strive.

Notes

1. Moore builds directly on Foster, but as I explore below, in the current acrid intellectual environment, this depiction of the genealogy of views itself is likely contentious. Yet, Moore has stressed repeatedly his effort as attempting to stand "on the shoulders of" Foster, most recently praising and attempting to "affirm [metabolic rift's] dialectical core" (Moore 2017a).
2. All three have held positions in sociology, but Moore's PhD dissertation was in geography, and he identifies as "an environmental historian and historical geographer" (see https://jasonwmoore.com/). He also rejects the label ecosocialist (personal communication).
3. After the publication of the Special Issue on ecologically unequal exchange in the *Journal of World-Systems Research* (Frey, Gellert, and Dahms 2017) in which we termed this a non-debate (Gellert et al. 2017), I received email from Moore in which he rejected labeling it a "non-debate" because of the implied equivalent responsibility for the lack of debate. After reviewing more of the publicly available comments from Foster, including especially his interview with Ian Angus (2016), I find myself in agreement with Moore about the one-sided ad hominem attacks on him. Moore's

criticisms of Foster, metabolic rift, and other environmentalists are sharp and pointed, but he has repeatedly praised Foster's contributions and, on a personal level, only complained of the lack of engagement by Foster et al. with his ideas. Foster (2016) does address the intellectual questions of monism and dualism that Moore is interested in, but as far as I know, there has been no public debate on the merits of their positions (as of January 2018). Increasingly, I observe that Foster and his colleagues engage in a kind of intellectual shunning and simply do not cite Moore's work, although Foster's (2016) article was an exception for its extended critique.

4. In full disclosure, I do not come completely unbiased into this discussion; Bunker was the chair of my PhD dissertation at the University of Wisconsin in 1998.
5. For a critique of Moore's concept of "exhaustion" from a soil scientist, see Engel-Di Mauro (2014).
6. The result is that it is difficult if not impossible to remain neutral or appreciative of the complexities and distinctions in this debate, as I try to do here. Good friends have cautioned me against stepping into the cow pies. As I am not a member of Facebook, I have not joined debates there, although there have been mentions of this in some of the more publicly available material.
7. Bunker (1985:99) also rejected the application of the label "frontier" to the Amazon because it incorrectly implied an eventual incorporation which he deemed unlikely, assumed an expansion into empty space rather than conflict between different systems, and implies linear progress while he saw discontinuous change, especially due to mining.
8. One silly issue between Foster and Moore is the former's displeasure with having the metabolic rift argument attributed to him, rather than Marx. Foster complains that he has been falsely accused of being the author of the idea of a metabolic rift denying Marx the credit. Marx had written of the "irreparable rift in the interdependent process of social metabolism," (Foster 2013) but given Foster's meticulous attention to the archives and his efforts to "rescue" Marx from the critiques of what he now calls the first wave of eco-Marxist thought, it seems a sort of false modesty to be chagrined by the accusation that the metabolic rift is "his." As Chew and Sarabia (2016:3) caustically observe, Foster has been "mining the seams of Marx's mother lode, especially *Das Kapital*, even to the level of footnotes to support [his] attribution that Marx had always paid attention to Nature in his writings."

9. In the interest of full disclosure, I wrote a laudatory review of Foster's earlier book on the Ecological Revolution in which I praised him for bringing empirical precision to claims of ecological crisis. At this point, however, I increasingly find reliance on the weight of scientific evidence of "natural" disaster to be distressingly apocalyptic and, although he surely does not intend to do so, leaves the possibility of readers detaching this diagnosis from the causes in capitalist accumulation that represent the power of eco-Marxist perspectives.
10. In a new article Moore (2017a) attributes this difference to Foster's exclusion of geographers and geography as a discipline.
11. At the Knoxville Conference there was excessive debate, in my view, over the question of whether particular attendees were "optimistic" or "pessimistic" about the future (see Killian 1971).

References

Angus, Ian. 2016. "In Defense of Ecological Marxism: John Bellamy Foster Responds to a Critic." *Climate & Capitalism*. Retrieved December 30, 2016 at http://climateandcapitalism.com/2016/06/06/in-defense-of-ecological-marxism-john-bellamy-foster-responds-to-a-critic/.

Arrighi, Giovanni. 1998. "Capitalism and the Modern World-System: Rethinking the Nondebates of the 1970s." *Review (Fernand Braudel Center)* XXI(1):113–129.

Bakker, Karen and Gavin Bridge. 2006. "Material Worlds? Resource Geographies and the 'Matter of Nature'." *Progress in Human Geography* 30(1):5–27.

Bunker, Stephen G. 1984. "Modes of Extraction, Unequal Exchange, and the Progressive Underdevelopment of an Extreme Periphery: The Brazilian Amazon, 1600–1980." *American Journal of Sociology* 89(5):1017–1064.

———. 1985. *Underdeveloping the Amazon: Extraction, Unequal Exchange, and the Failure of the Modern State*. Chicago, IL: The University of Chicago Press.

———. 2005. "How Ecologically Uneven Developments Put the Spin on the Treadmill of Production." *Organization & Environment* 18(1):38–54.

———. 2006. *The Snake with Golden Braids: Society, Nature, and Technology in Andean Irrigation*. New York: Lexington Books.

Bunker, Stephen G. and Paul S. Ciccantell. 2003. "Generative Sectors and the New Historical Materialism: Economic Ascent and the Cumulatively Sequential Restructuring of the World-Economy." *Studies in Comparative International Development* 37(4):3–30.

———. 2005. *Globalization and the Race for Resources*. Baltimore, MD: The Johns Hopkins University Press.

Chew, Sing C. and Daniel Sarabia. 2016. "Nature–Culture Relations: Early Globalization, Climate Changes, and System Crisis." *Sustainability* 8(78):1–29.

Clark, Brett and John Bellamy Foster. 2012. "Guano: The Global Metabolic Rift and the Fertilizer Trade." Pp. 68–82 in *Ecology and Power; Struggles over Land and Material Resources in the Past, Present and Future*, edited by Alf Hornborg, Brett Clark, and Kenneth Hermele. New York: Routledge.

Ecomodernist Manifesto. 2015. "An Ecomodernist Manifesto" (by John Asafu-Adjaye, Linus Blomqvist, Stewart Brand, et al.). Retrieved June 14, 2016 at http://www.ecomodernism.org/.

Engel-Di Mauro, Salvatore. 2014. *Ecology, Soils, and the Left: An Eco-Social Approach*. New York: Palgrave Macmillan.

Foster, John Bellamy. 1999. "Marx's Theory of Metabolic Rift: Classical Foundations for Environmental Sociology." *American Journal of Sociology* 105(2):366–405.

———. 2000. *Marx's Ecology*. New York: Monthly Review Press.

———. 2009. *The Ecological Revolution: Making Peace with the Planet*. New York: Monthly Review Press.

———. 2013. "Review of the Month: Marx and the Rift in the Universal Metabolism of Nature." *Monthly Review* 65(7). Retrieved August 31, 2017 at https://monthlyreview.org.

———. 2016. "Marxism in the Anthropocene: Dialectical Rifts on the Left." *International Critical Thought* 6(3):393–421.

———. 2017. "The Earth-System Crisis and Ecological Civilization: A Marxian View." *International Critical Thought* :1–20. https://doi.org/10.1080/21598282.2017.1357483.

Foster, John Bellamy and Brett Clark. 2012. "Review of the Month: The Planetary Emergency." *Monthly Review* 64(7). Retrieved August 31, 2017 at https://monthlyreview.org.

———. 2016. "Review of the Month: Marx's Ecology and the Left." *Monthly Review* 68(2). Retrieved August 31, 2017 at https://monthlyreview.org/2016/06/01/marxs-ecology-and-the-left/.

Foster, John Bellamy and Fred Magdoff. 2011. *What Every Environmentalist Needs to Know About Capitalism*. New York: Monthly Review Press.

Foster, John Bellamy and Hannah Holleman. 2014. "The Theory of Unequal Ecological Exchange: A Marx-Odum Dialectic." *The Journal of Peasant Studies* 41(2):199–233.

Foster, John Bellamy, Brett Clark, and Richard York. 2010. *The Ecological Rift: Capitalism's War on the Earth*. New York: Monthly Review Press.

Freudenburg, William R., Scott Frickel, and Robert Gramling. 1995. "Beyond the Nature/Society Divide: Learning to Think About a Mountain." *Sociological Forum* 10(3):361–392.

Frey, R. Scott, Paul K. Gellert, and Harry F. Dahms, Guest Editors. 2017. Special Issue on Unequal Ecological Exchange. *Journal of World-Systems Research* 23(2):226–398. Retrieved November 30, 2017 at http://jwsr.pitt.edu/ojs/index.php/jwsr/article/view/733.

Gellert, Paul K. 2005. "For a Sociology of 'Socionature': Ontology and the Commodity-Based Approach." Pp. 65–91 in *Nature, Raw Materials, and Political Economy, Research in Rural Sociology and Development*, edited by P. S. Ciccantell, G. Seidman and D. A. Smith. Amsterdam, The Netherlands: Elsevier JAI.

Gellert, Paul K., R. Scott Frey, and Harry F. Dahms. 2017. "Introduction to Ecologically Unequal Exchange in Comparative Perspective." *Journal of World-Systems Research* 23(2):226–235. Retrieved November 30, 2017 at http://jwsr.pitt.edu/ojs/index.php/jwsr/article/view/733.

Goldman, Michael and Rachel A. Schurman. 2000. "Closing the 'Great Divide': New Social Theory on Society and Nature." *Annual Review of Sociology* 26:563–584.

Holleman, Hannah. 2015. "Method in Ecological Marxism: Science and the Struggle for Change." *Monthly Review* 67(5). Retrieved November 30, 2017 at https://monthlyreview.org/2015/10/01/method-in-ecological-marxism/.

Kallis, Giorgos and Richard B. Norgaard. 2010. "Coevolutionary Ecological Economics." *Ecological Economics* 69(4):690–699.

Killian, Lewis M. 1971. "Optimism and Pessimism in Sociological Analysis." *The American Sociologist* 6(4):281–286.

Liu, Jianguo, Thomas Dietz, Stephen R. Carpenter, Marina Alberti, Carl Folke, Emilio Moran, Alice N. Pell, Peter Deadman, Timothy Kratz, Jane Lubchenko, Elinor Ostrom, Zhiyun Ouyang, Wiliam Provencher, Charles L. Redman, Stephen H. Schneider, and William W. Taylor. 2007. "Complexity of Coupled Human and Natural Systems." *Science* 317:1513–1516.

Longo, Stefano B., Rebecca Clausen, and Brett Clark. 2015. *The Tragedy of the Commodity: Oceans, Fisheries, and Aquaculture*. Rutgers, NJ: Rutgers University Press.

Moore, Jason W. 2000. "Sugar and the Expansion of the Early Modern World-Economy: Commodity Frontiers, Ecological Transformation and Industrialization." *Review (Fernand Braudel Center)* 23(3):409–433.

———. 2003. "Capitalism as World-Ecology. Braudel and Marx on Environmental History." *Organization & Environment* 16(4):431–458.

———. 2009. "Madeira, Sugar, & the Conquest of Nature in the 'First' Sixteenth Century, Part I: From 'Island of Timber' to Sugar Revolution, 1420–1506." *Review (Fernand Braudel Center)* 32(4):345–390.

———. 2010a. "Madeira, Sugar, & the Conquest of Nature in the 'First' Sixteenth Century, Part II: From Local Crisis to Commodity Frontier, 1506–1530." *Review (Fernand Braudel Center)* 33(1):1–24.

———. 2010b. "'Amsterdam is Standing on Norway' Part I: The Alchemy of Capital, Empire and Nature in the Diaspora of Silver, 1545–1648." *Journal of Agrarian Change* 10(1):33–68.

———. 2011a. "Transcending the Metabolic Rift." *Journal of Peasant Studies* 38(1):1–46.

———. 2011b. "Ecology, Capital and the Nature of Our Times: Accumulation & Crisis in the Capitalist World-Ecology." *Journal of World-Systems Research* 17(1):109–147.

———. 2014. "Toward a Singular Metabolism: Epistemic Rifts and Environment-Making in the Capitalist World-Ecology." *New Geographies* 6:10–19.

———. 2015a. *Capitalism in the Web of Life: Ecology and the Accumulation of Capital.* New York: Verso.

———. 2015b. "Cheap Food and Bad Climate." *Critical Historical Studies* 2(1):1–43.

———, editor. 2016. *Anthropocene or Capitalocene? Nature, History, and the Crisis of Capitalism.* Oakland, CA: PM Press.

———. 2017a. "Metabolic Rift or Metabolic Shift? Dialectics, Nature, and the World-Historical Method." *Theory and Society*. https://doi.org/10.1007/s11186-017-9290-6.

———. 2017b. "The Capitalocene, Part I: On the Nature and Origins of Our Ecological Crisis." *The Journal of Peasant Studies* 44(3):594–630.

———. 2017c. "The Capitalocene, Part II: Accumulation by Appropriation and the Centrality of Unpaid Work/Energy." *The Journal of Peasant Studies* :1–43. https://doi.org/10.1080/03066150.2016.1272587.

Norgaard, Richard. 1994. *Development Betrayed: The End of Progress and a Co-Evolutionary Revisioning of the Future.* New York: Routledge.

Ollman, Bertell. 2003. *Dance of the Dialectic: Steps in Marx's Method.* Champaign, IL: University of Illinois Press.

O'Connor, James. 1988. "Capitalism, Nature, Socialism: A Theoretical Introduction." *Capitalism Nature Socialism* 1(1):11–38.

O'Hearn, Denis. 2007. "Bringing the Human Back into the Material: Embodied Perception in Stephen Bunker's Political Economy." *Review (Fernand Braudel Center)* 30(2):145–160.

Salleh, Ariel. 2005. "Moving to an Embodied Materialism." *Capitalism Nature Socialism* 16(2):9–14.

Schneider, Mindi and Philip McMichael. 2010. "Deepening, and Repairing, the Metabolic Rift." *The Journal of Peasant Studies* 37(3):461–484.

Swyngedouw, Erik. 1999. "Modernity and Hybridity: Nature, *Regeneracionismo*, and the Production of the Spanish Waterscape, 1890–1930." *Annals of the Association of American Geographers* 89(3):443–465.

White, Damian F., Alan P. Rudy, and Brian J. Gareau. 2015. *Environments, Natures and Social Theory: Towards a Critical Hybridity*. New York: Palgrave Macmillan.

York, Richard and Riley E. Dunlap. 2012. "Environmental Sociology." Pp. 504–521 in *The Wiley-Blackwell Companion to Sociology*, edited by G. Ritzer. Malden, MA: Blackwell.

Part II

Cases of Ecologically Unequal Exchange in Comparative and Historical Context

6

The Entropy Curse

Laura McKinney

The exacerbating effects of economic development and ecologically unequal exchanges on various facets of environmental degradation are well documented (Bunker 1985; Frey 1998; Jorgenson and Clark 2012). What is less explored in the literature is the inverse, that is, the influence of environmental decline on development outcomes, which is precisely what this chapter provides. The chapter advances a framework for understanding the ways in which environmental degradation undercuts economic development in less-developed countries. To do so, I draw on existing theoretical and empirical work in the ecologically unequal exchange tradition (e.g., Bunker 1985; Hornborg 2001; Jorgenson and Rice 2007; Lawrence 2009), with specific focus on the biophysical basis of underdevelopment resulting from the depletion of natural resources. I also advance an ecologically informed lens for understanding the empirical findings that form the foundation of the resource curse (i.e., Sachs and Warner 1995, 2001). In doing so, the chapter demonstrates that the

L. McKinney (✉)
Department of Sociology, Tulane University, New Orleans, LA, USA
e-mail: lauramc@tulane.edu

R. S. Frey et al. (eds.), *Ecologically Unequal Exchange*,
https://doi.org/10.1007/978-3-319-89740-0_6

natural resource curse literature supports the basic tenets of the ecologically unequal exchange tradition and kindred perspectives, namely world-system analysis (e.g., Wallerstein 1974, 2004) and dependency frameworks (e.g., Frank 1978).

The natural resource curse hypothesis—as treated by various scholars in economics and political science, among other disciplines—is a widely adhered to explanation of development differentials witnessed the world over (see, e.g., Auty 2001; Frankel 2010; Rajan 2011; Ross 1999; Rosser 2006; Sachs and Warner 1995, 2001). The hypothesis contends that resource-abundant nations have experienced poor performance on numerous indicators of development, though the bulk of the empirical evidence centers on measures of economic growth. This position is nicely encapsulated by remarks of leading resource curse proponents such as Auty (2001:840) who states, "Since the 1960s, the resource-poor countries have outperformed the resource-rich countries compared by a considerable margin," and Sachs and Warner (1995:2) who assert, "one of the surprising features of modern economic growth is that economies abundant in natural resources have tended to grow slower than economies without substantial natural resources." This body of work is often invoked to account for the relatively low levels of development observed in peripheral nations that are characteristically rural and resource rich. This chapter critiques the natural resource curse literature on conceptual and analytic grounds, ultimately concluding that "resource curse" findings provide strong support for the theoretical predictions of global political economy perspectives in environmental sociology, particularly those in the ecologically unequal exchange tradition.

In what follows I treat the theoretical roots of the ecologically unequal exchange tradition with specific emphasis on the thermodynamics of underdevelopment and elaborate the conceptual and analytic critiques of the natural resource curse literature. Incorporating the laws of thermodynamics, in particular the law of entropy, leads to an ecologically informed understanding of the empirical findings springing from the natural resource curse literature that, when seen in this light, provides a bevy of support for ecologically unequal exchange theorizations. The chapter draws on physical science and thermodynamic principles to substantiate further the central claim that it is the *liquidation* of resources—not resource abundance per se—that stunts economic growth in the periphery. In essence, the structure of the world-system and ecologically unequal exchanges therein fuel the appropriation of resources that stunts development in less-developed nations. An analysis of the drivers of economic development in less-developed countries yields empirical support for the theoretical tenets of ecologically unequal exchange.

Ecologically Unequal Exchange and the Entropy Curse

Ecologically unequal exchange research demonstrates that energy and natural resources disproportionately flow from poor to rich countries, which is a maxim adhered to by a number of kindred perspectives. For instance, the unifying threads of ecologically unequal exchange perspectives, world-system analysis (Wallerstein 1974, 2004), dependency theory (Bunker 1985; Frank 1978), and metabolic rift analyses (Foster 1999; Foster, Clark, and York 2010)—and the potential for a comprehensive theory of unequal ecological exchanges springing from their synthesis—have been documented (Foster and Holleman 2014; Moore 2000). The premise shared by these perspectives and the foundational work in which they are rooted (e.g., Baran 1957; Grossmann 1929; Marx 1867/1977; Prebisch 1950; Singer 1950) is that exchanges in the world-system consistently and disproportionately favor core nations that reap economic, political, social, and environmental advantages while conferring to peripheral locales environmental degradation, economic dependency, political instability, and underdevelopment, broadly defined. Of primary interest is the appropriation of energy, natural resources, and raw materials from peripheral areas to core nations that constitute the ecologically unequal exchanges that perpetuate global inequality.

Within this tradition, empirical analyses confirm the adverse environmental outcomes for less-developed countries emanating from ecologically unequal exchanges in the world-system, which range from deforestation (Shandra, Leckband, and London 2009a) and methane emissions (Jorgenson 2006) to threats to biodiversity (Shandra et al. 2009b) and sustainability losses (Rice 2007). In addition, empirical analyses confirm that economic development exacerbates environmental losses and is at odds with global sustainability (e.g., McKinney 2012; York, Rosa, and Dietz 2003). These and myriad other forms of environmental degradation commonly examined by sociologists are symptomatic of general ecological disorder indicative of increased entropy. Despite their overarching importance and treatment in other disciplines (e.g., economics, see Georgescu-Roegen 1971; anthropology, see Hornborg 2001; human ecology, see Rees 2004; ecological economics, see Daly and Cobb 1989), most theories in environmental sociology have yet to fully embrace the entropic dynamics that anchor nearly all ecological phenomena of interest. What is missing, and what this chapter offers, is a theoretical and empirical approach to environmental degradation that accounts for the underlying element of deterioration of the physical world—the generation of entropy.

Entropy is a foundational concept in the physical sciences that is treated as a fundamental law of nature, as explained by the second law of thermodynamics (Georgescu-Roegen 1971; Prigogine 1967). In the physical world, degradation of order into chaos is an irreversible and continuous process that is captured by the term "entropy" and is intrinsically linked to issues of environmental sustainability (Daly and Cobb 1989). While the first law of thermodynamics states that no energy may be created or destroyed (which could be taken to suggest limitless energy usage), it is the second law of thermodynamics that captures what is lost—the "orderliness" or quality of energy that is irreversibly impacted when applied in production. Essentially, entropy refers to the capacity for rearrangement; as an entity tends toward high entropy, it is less able to be rearranged and thus less useful (Prigogine 1967). Pristine natural resources exist in states of low-entropy with maximum capacity for rearrangement so that "work" may be done, but processes of extraction and production reconfigure raw materials, create pollution, and generate entropy or disorder.

Low-entropy natural resources range from air, land, water, coal, oil, wood, plant, and animal biota and serve multiple purposes related to the current sustainability of all life forms. Economic advance in core nations is accelerating the deterioration of the physical world due to its dependence on natural resource inputs and the pollution it creates

(Georgescu-Roegen 1971). Importantly, the structure of ecologically unequal exchange in the world-system concentrates the generation of entropy in peripheral locales while preserving environmental order in the core (Biel 2006; Hornborg 2001, 2015). Thus, entropic dynamics are inextricably linked to economic processes and the social relations in which they are embedded.

The interest in entropy has deep roots in the social sciences. To illustrate, Foster (1999:370) and colleagues (Foster and Holleman 2012) highlight Weber's critique of Nobel prize recipient Wilhelm Ostwald's theory of social energetics, which posits that the entropy law applies to both energy and matter (Ostwald neglected the latter). Burkett and Foster (2006) offer a compelling chronology of written exchanges between Marx, Engels, and physicists at the forefront of research on what later came to be known as the laws of thermodynamics to substantiate the claim that entropic dynamics are fully accounted for in the theory of capital (but see Hornborg 2015). In economics, the entropy law is applied by Georgescu-Roegen (1971) to reject the Ricardian land assumption (once fully embraced by the discipline) that nature provides free gifts, the flow of which continues in perpetuity without declines in abundance or vitality, which is in clear contradiction to the basic laws of thermodynamics (see also Daly and Cobb 1989). Despite the significance identified by classical theorizations that anticipate the chief importance of resource scarcity as it relates to the availability of quality energy and matter, entropic dynamics represent a relatively under-utilized component in the sociological discipline, though there are some exceptions, discussed below.

In addition to the classical theorists, foundational treatments in the subfield of environmental sociology emphasize the role of biophysical constraints and natural resource limits to socioeconomic conditions (Catton 1980). As suggested by some theoreticians working in the ecologically unequal exchange tradition and international political economy, more generally, entropic disorder accompanying the extraction and appropriation of natural resources in the periphery for export to the core is a key axis perpetuating current patterns of global inequality (Biel 2006; Hornborg 2001, 2014). Biel (2006) notes that characteristics of the current system of capital are the decidedly unsustainable production and exchange processes that undermine our ecological base, with the greatest assaults to sustainability concentrated in peripheral locales (see also Rice 2007). Frank (2005) notes that the dissipation of entropy (disorder) is integral to dependency dynamics, although, unfortunately, he did not fully develop this line of argumentation. The historical advocacy of interdisciplinary perspectives notwithstanding, current approaches in sociology, generally, and environmental sociology, in particular, lack analytical emphasis on the transformation of our biophysical environment (i.e., energy and matter) as both a cause and effect of social and economic arrangements (Schultz and York 2011).[1] This chapter modestly contributes to this gap by incorporating entropic dynamics into key theorizations in environmental sociology to interrogate environmental degradation as a cause of underdevelopment.

The ecologically unequal exchange approach offers that entropy also shapes domestic social factors, predicting equally important negative social impacts for the poorer countries involved in such exchanges. For example, Biel (2006) persuasively argues that social order in the core comes at the expense of social "disorder" in peripheral populations. Drawing on the concept of core and periphery relations, the system of capital necessarily generates excess entropy that must be absorbed by subordinate actors on the North–South divide. The depletion of resources severely threatens sustainability and can result in devastating degradation that leads to the displacement of entire communities and the emergence of state-led repression that violates basic human rights (Bonds and Downey 2012). These predictions parallel those emerging from the resource curse literature, to which I now turn.

The Resource Curse: A Conceptual and Analytic Critique

An important supplement to advancing the framework above centers on critiquing the resource curse literature's argument that resource abundance is a primary causal agent that stunts development in poor nations (see, e.g., Frankel 2010; Rajan 2011; Ross 1999; Rosser 2006; Sachs and Warner 1995, 2001). I critique the natural resource curse literature on the following conceptual and analytic grounds: (1) failing to acknowledge that all socioeconomic progress depends on environmental inputs; (2) the empirical bases upon which the hypothesis is purported to find support; and (3) the implications of key resource curse findings that strongly support the theoretical predictions of global political economy perspectives in environmental sociology, including the ecologically unequal exchange tradition. Incorporating the entropy law further substantiates the central claim of the chapter that it is the *liquidation* of resources—not resource abundance per se—that stunts economic growth in the periphery. What follows is an elaboration of these central critiques and discussion of the principal studies that constitute the empirical backbone of the resource curse tradition.

Conceptually, the resource curse hypothesis is at odds with an ecologically informed position that acknowledges the life-sustaining functions ecosystems provide to all of humanity. Quite simply, all social and economic progress depends on resources garnered from the environment (see, e.g., Crosby 2004; Diamond 1997; McNeill 2001). Ecosystem functions are fundamental to societal advance, economic progress, and individual well-being. Resources garnered from the environment provide necessities for human life such as food, clothing, and shelter. Harnessing energy and harvesting resources are essential components of production processes that contribute to economic advance. Without a hospitable and supportive environment, societies cannot exist and economies cannot operate. Thus, the assertion that resource abundance portends a curse for societies suggests a general failure to acknowledge the ways in which human societies benefit from natural resources.

Analytically, the bulk of the empirical evidence upon which the resource curse literature is purported to find support belies the sweeping generalization that resource abundance confers developmental losses. Sachs and Warner's (1995) extremely influential publication, "Natural Resource Abundance and Economic Growth," laid the empirical foundation for the resource curse hypothesis. The major conclusion imparted by the analysis is the depressing effect of resource abundance on development, which informs their conclusion of a resource curse. One assumption of the analyses that constitute empirical support for the resource curse hypothesis is that the ratio of primary exports to domestic production is an acceptable proxy for resource abundance (Saad-Filho and Weeks 2013; Stijns 2005; Wright and Czelusta 2004). However, export ratios reflect a wide range of domestic factors and "export composition measures say little about the availability of resources" (De Soysa 2000:122). Some resource-rich countries, such as the United States, do not exhibit high primary export ratios, but to conclude a lack of resource wealth on this basis would be fallacious. What Sachs and Warner are measuring is resource *dependency*, or the degree to which domestic production and trade center on primary sector commodities (Ding and Field 2005; Saad-Filho and Weeks 2013; Stijns 2005). In this light, the resource curse literature shows remarkable consistency with and substantial empirical support for the theoretical predictions of ecologically unequal exchange perspectives as well as the dependency and world-system perspectives.

The evidence that Sachs and Warner (1995, 2001) deem indicative of a resource curse is precisely the foundation upon which ecologically unequal exchange and kindred perspectives lie. Beginning with the work of Singer (1950) and Prebisch (1950), and continuing with Frank (1978) and Amin (1974), the developmental adversities for peripheral countries engaged in the exchange of primary products for the finished products of richer countries that benefit from this arrangement has been established as an historical feature of the global hierarchy of nations, the global division of labor, and ecologically unequal exchanges therein. Notably, this perspective has remarkable symmetry with Marx's position that capitalist industrial countries "convert one part of the globe into a chiefly agricultural field of production for supplying the other part" (Marx 1867/1977:579–580). Further theoretical and empirical refinements in this vein have elaborated the ways in which pricing systems are at odds with thermodynamic principles (specifically, the entropy law), and essentially reward the dissipation of entropy by assigning higher market values to (more entropic) manufactured goods compared to the (low-entropy) raw material inputs (Biel 2006; Hornborg 2001). Thus, it is unsurprising that the depletion of low-entropy natural resources, the supply of which we all depend on, stunts social progress and economic growth in poor countries whose economies are dominated by primary sector production and exchange.

Another area of potential overlap between the resource curse and ecologically unequal exchange perspectives is the importance of political systems to development. Theoretically speaking, the resource curse is articulated as a set, or a package, of unfavorable social, political, and economic conditions associated with the abundance of resources (Karl 1997, 2004; Stiglitz 2007). There is extensive empirical support that extractive economies distort the development of democracy, in general, and fuel adverse political conditions of unstable governance, authoritarian regimes, poor policy, a lack of transparency and other related institutional failures that some posit are more deleterious to development than the direct effect of resource booms (e.g., Collier and Hoeffler 2005; Dunning 2005; Polterovich, Popov, and Tonis 2007; Robinson, Torvik, and Verdier 2006; Ross 2001). Similarly, ecologically unequal exchange and dependency scholars note the bevy of adversities associated with extractive economies, including negative effects on democracy (Bollen 1983; Bunker 1985). Taken together with the framework advanced above, we can expect resource losses to impede the installation and stability of democratic regimes.

In order to assess empirically the central claim of this article that the liquidation of natural resources is a key agent perpetuating global social and economic inequality, I offer a cross-national analysis to test the dynamics of interest. To anticipate, accumulated ecological losses are found to be at odds with economic development, net of relevant controls. I now turn to a discussion of the data and method employed.

Data and Methods

Presented below is an analysis of economic growth for the period 2002–2012 to assess the theoretical propositions outlined above. Consistent with prior research, cases are restricted to 115 less-developed nations. I construct a structural equation model (SEM) to analyze the drivers of economic development, given its favorable treatment of error terms[2] and the incorporation of composite indicators that decreases potential bias of estimates for models with highly intercorrelated independent variables. Moreover, SEMs provide model fit statistics that enable the researcher to judge the fit of the model as a whole to the data provided, and make adjustments based on this information (Bollen 1989). Another favorable feature of an SEM is its use of maximum likelihood estimates (MLE) that calculate pathway coefficients on the basis of all available data points; when cases are missing information on select variables those cases are dropped from the pathway estimations but retained for others when the data are available. Thus, an SEM maximizes the sample of nations, which is a crucial concern for relatively small samples.

The research design utilizes a time-ordered dependent variable, where the dependent variable is measured in time after the independent variables. This strategy is commonly used in cross-sectional macro-comparative research in order to help adhere to conditions of causality (e.g., McKinney 2014; Shircliff and Shandra 2011). Also, substantively, it is likely that the effects of environmental decline on economic development accumulate over time; that is, resource depletion that inhibits economic advance may take years to manifest. Thus, biocapacity losses are measured over a fairly significant period of time (1971–2001), and the final outcome of growth in GDP/c is measured for the decade of 2002–2012. Other relevant controls are included, as explained below.

Variables Included in the Analysis

Growth in GDP/c is the ultimate dependent variable and indicates net change in each nation for the period 2002–2012. Although GDP/c has been heavily critiqued as an indicator of development, it is retained in the analysis given empirical precedent as a favored outcome of interest in empirical analyses in the natural resource curse (Sachs and Warner 1995, 2001) and world-system tradition (e.g., Bornschier, Chase-Dunn, and Rubinson 1978; Chase-Dunn 1975; Dixon and Boswell 1996; Kentor 2001). As established in these analyses, initial values of *GDP/c in 2002* are included as a regressor in order to stabilize comparisons of growth for economies with relatively larger or comparatively smaller bases. GDP per capita is the sum of value added by all resident producers in a country in a given year, divided by the midyear population. It is reported in constant dollars taken from the World Bank (2013). Change scores are calculated using the typical formula—$(T_2 - T_1)/T_1 \times 100$—wherein the difference in values for 2012 and 2002 are divided by initial values at time one, then transformed into a percentage.

Biocapacity loss is a key independent variable in this analysis, measured as the percent change in domestic ecological resources for the years 1971–2001.[3] This variable, taken from the Global Footprint Network (2010), quantifies reductions in the amount of biologically productive resources available to individuals in a nation and is comprised of stocks of grazing land, cropland, forestland, and fishing grounds. Although select indicators of natural capital losses could be employed, this composite variable is a preferred measure given the comprehensive assessment of diverse indicators of environmental destruction that closely depicts the hypotheses outlined above. Another feature is that it is a per capita measure, which provides a meaningful basis for comparison that illuminates the severity of ecological crises and resource declines as distributed across a given population. Biocapacity loss is taken as an exogenous variable, co-varied with other exogenous terms, and specified as directly and indirectly influencing economic growth. Expectations based on the literature reviewed above are that losses are the primary cause of poor economic growth.

In addition to the initial values of GDP/c, other relevant controls and significant predictors of economic growth commonly employed in natural resource curse empirical tests (Sachs and Warner 1995) are included in the analysis. Specifically, *percent rural* and *gross capital formation* for the year 2001 (World Bank 2013) are included given the proclivity for rural areas to be targeted as sites for resource extraction and the potential for domestic investment to boost economic growth, respectively. In addition, a latent variable comprised of *primary and secondary school enrollment* rates (World Bank 2013) is included based on the general tenet that advancing human capital via educational attainment is a fruitful avenue for generating economic growth. *Primary exports as a percent of GDP* is included as an indicator of "resource abundance" because it has been used in previous analyses of the resource curse, and primary exports are intimately linked to environmental losses. Primary export ratios are for the year 2001 (World Bank 2013); this variable is specified as mediating the relationship of percent rural areas and biocapacity losses to the outcome.

FDI stock accumulation 1990–2000 is also included, which represents the net change in FDI stock as a percent of GDP from 1990 to 2000 (calculated as $T_2 - T_1$). Foreign investment data are taken from UNCTAD (2014).

Analysis and Discussion of Results

Before interpreting the findings, it is important to note the limitations of this study. First, the sample consists of all less-developed countries, though regional variations might exist and should be explored. The research design prohibits this kind of analysis, as an SEM is not well equipped to handle dummy variables that could otherwise delineate unique regional dynamics. Alternatively, separate analyses could be run on each regional group of nations so long as a sufficient case base is reached through the application of, for instance, time-series data and models. Though this falls outside the scope of the current study, such avenues for future research are highly encouraged. A related weakness is the inability to account for potential heterogeneity bias in cross-sectional models, which certain longitudinal methods could address.[4] Finally, the analysis is inclusive of select measures of dependency. However, the structure of domestic trade (e.g., narrowness of export partners and traded commodities) is yet another dimension of dependency that future research could explore.

As an initial first step in the analysis, descriptive statistics and bivariate correlations were examined. With regard to the latter, the magnitude of the relationships among the variables demonstrates that many of the predictor variables are highly correlated, which further warrants the use of an SEM given its superior handling of intercorrelated independent variables through the creation of latent constructs and direct and indirect pathways that circumvent the tendency to bias coefficient estimates (e.g., Bollen 1989; Byrne 2009). Before presenting the results, it is worth noting that political indicators of regime type (a ten-point scale ranging from autocracy to democracy), civil liberties, and political rights were included in initial model estimations but failed in all cases to reach statistical significance and worsened model fit.[5]

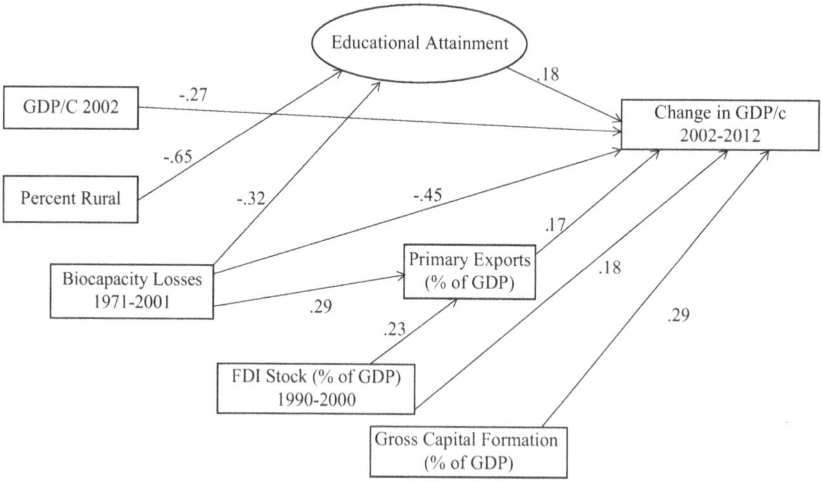

Fig. 6.1 Structural equation model predicting change in GDP per capita 2002–2012 in less-developed nations (*N* = 115)

Figure 6.1 presents the final SEM results of growth in GDP/c for 2002–2012 in less-developed nations as directly and indirectly conditioned by the environmental, social, and economic factors treated above. The model was derived through an iterative process of testing all theoretically and substantively informed paths proceeded by successive steps of eliminating non-significant relationship(s) and model re-estimation to maximize parsimony and optimize model fit, which is standard practice in the SEM tradition (Byrne 2009). The final model evidences an excellent fit to the data in accordance to empirical standards: a non-significant chi-square (χ^2 = 15.620, df = 14, p = 0.337), low RMSEA (0.032) value well below the suggested upper threshold of 0.05, and IFI (0.995) and CFI (0.994) values that exceed minimum acceptable value of 0.90 (Bollen 1989). The satisfactory fit statistics permit interpretation of coefficients, to which I now turn.

The results confirm that biocapacity loss represents an important underlying factor that directly and indirectly shapes trajectories of economic development in poor countries. In fact, the magnitude of the direct effect of biocapacity loss on change in GDP/c is unrivaled in the results and demonstrates the sizeable diminishing effect (−0.45) of environmental losses on economic development. Environmental losses indirectly influence change in GDP/c via their depressing effect on education (−0.32), which in turn positively effects (0.18) economic growth. Thus, environmental losses negatively impact development trajectories in direct and indirect ways. Results also confirm that primary exports are predicted by ecological losses (0.29) and FDI stock accumulation (0.23). Interestingly, when the effect of biocapacity loss is accounted for the comparatively small but favorable impact of primary exports (0.17) and FDI stock accumulation (0.18) on economic growth is evident, but the negative effect of biocapacity losses on development outstrips potential gains. Collectively, the results regarding the influence of environmental losses are in line with the hypotheses derived from an ecologically informed analysis of development and provide support for the basic premise of this chapter that it is the liquidation of natural resources (not their presence) that stymies development.

The prevalence of rural areas is found to have a strong, negative association with educational attainment (−0.65), which is subsequently tied to economic growth (0.18). As expected, domestic investment yields positive effects (0.29) on change in GDP/c. The standardized coefficients demonstrate that the biggest boost to economic growth comes from domestic investment, suggesting that initiatives to incite internal capital formation are powerful avenues for advancing poor countries' development trajectories. Secondly, the positive effect of educational attainment suggests that investments in boosting human capital in a nation are also advantageous to spurring economic growth. In sum, the results indicate that strategies to foster domestic investments and cultivate human capital are efficacious for ten-year economic growth. Initial values of GDP/c in 2002 support the basic notion that larger bases tend to exhibit slower growth (−0.27), which is in line with prior empirical results.

Comparing the standardized regression coefficients in Table 6.1, we find that biocapacity losses has the largest total influence on change in GDP/c (−0.462), which accounts for direct and indirect effects. Gross capital formation has the next largest total effect on the outcome (0.286), closely followed by initial values of GDP/c (−0.273). FDI stock accumulation and primary exports are found to have positive total effects on GDP/c (0.225 and 0.172, respectively), once the depressing effect of

Table 6.1 Direct, indirect, and total effects on change in GDP/c 2002–2012

	Direct effects	Indirect effects	Total effects
FDI stock accumulation	0.185* (0.369)	0.040** (0.081)	0.225** (0.450)
Percent rural		−0.118*** (−0.197)	−0.118*** (−0.197)
Biocapacity losses	−0.452*** (−1.018)	−0.009** (−0.021)	−0.462*** (−1.039)
Primary exports	0.172* (665.5)		0.172* (665.5)
Gross capital formation	0.286*** (1.457)		0.286*** (1.457)
GDP per capita	−0.273** (−8.60)		−0.273*** (−8.60)
Education	0.183⁺ (0.599)		0.183⁺ (0.599)

⁺$p < 0.10$; *$p < 0.05$; **$p < 0.01$; ***$p < 0.001$
Standardized coefficients reported; unstandardized coefficients are in parentheses

ecological losses is accounted for. These results suggest that, in line with the arguments put forth above, it is the liquidation of natural resources that matters for development. Importantly, the marginal positive effects of FDI accumulation and primary exports combined do not offset the massive detrimental effect of biocapacity losses on growth. The results presented illustrate that although some variables provide only direct or indirect effects, examining the total effects gives a broader picture as to which factors explain the most variation in economic growth.

Concluding Remarks

The overarching goal of this chapter has been to theoretically develop and empirically analyze the effects of environmental degradation on economic performance in less-developed countries. The primary conclusion of the analysis is that environmental losses are extremely prohibitive for development in poor nations. This is an important point of departure from the resource curse literature that asserts resource-rich countries suffer poor developmental trajectories due to the "curse" of natural resources—that resource abundance creates unfavorable conditions for development.

This chapter develops and empirically supports that it is not the presence of natural resources that is prohibitive for development, but it is their liquidation that poses challenges.

Importantly, entropy is a concept taken from the physical sciences that encapsulates the explanatory mechanisms for the dynamics of interest. All natural resources exist in states of low-entropy that support all forms of life. As low-entropy natural resources are used and applied in production, they become less ordered and less useful. Thus, the degree to which abundant resources are liquidated presents serious challenges for development. Although many measures of development are available, the analysis predicts growth in GDP/c over a ten-year period (2002–2012) to substantiate these claims. The use of GDP/c as the dependent variable is an important limitation with implications for future analyses. In particular, I strongly advocate subsequent work to analyze the effect of environmental losses on a wide range of development outcomes, ranging from human/social development to quality of life indicators to political arrangements and the like. The analysis presented here is one small step in a long list of potential outcomes to be tested; hopefully it lays the groundwork for further research in this vein.

The results are provisionally suggestive of policy implications for poor nations seeking to improve conditions. Put simply, sustainable development that gives equal weight to the importance of environmental, economic, and social priorities is preferable. The results demonstrate that cumulative environmental losses are extremely prohibitive for future trajectories of growth, thus highlighting the advantages of adopting policies and practices to preserve natural resources as a means to foster long-term, sustainable growth. Similarly, the results support that investments in human capital (via educational attainment) and domestic capital formation are fruitful avenues for improving economic growth trajectories.

Collectively, the findings support the theoretical positions of ecologically unequal exchange traditions and broader, emerging themes in sustainability studies that elaborate the interconnected nature of the biophysical and social world (Liu et al. 2007). Of particular significance are those developments in the area of coupled human and natural systems (CHN) (Liu et al. 2007) that articulate the complex interactions of nature and society that include reciprocities, feedback loops, and telecoupling processes (see Liu et al. 2013) that shape the sustainability of local and distant places. In an increasingly globalized world, refinements in CHN analyses evidence a growing emphasis on the flows of materials and energy across sending and receiving systems, and the spillover effects they produce. Most important to the present effort, CHN analyses offer a framework for simultaneously considering socioeconomic and environmental interactions spanning systems from the local to the global level. These innovations and the theoretical framework presented here point to the increasing importance of myriad facets of global integration to sustainability outcomes across environmental, social, and economic dimensions.

Though many existing empirical analyses within the ecologically unequal exchange literature demonstrate the negative effects of globalized economies and world market exchanges on the environment, few explore the reverse. This chapter makes a modest contribution to filling that gap by confirming the substantial negative effect of environmental losses on subsequent economic growth in poor nations. The conclusions reached through the theory and analysis presented imply that there are large gains to be made by continuing research on myriad (human, social, health, economic, political) consequences of environmental losses. Emerging work in this vein suggests its importance (e.g., McKinney and Austin 2015) and justifies further interrogation of these crucial connections.

Notes

1. For exceptions, see Dietz and Jorgenson's (2013) edited collection on structural human ecology; see also Moore (2015) for a powerful interdisciplinary treatment of the history of capitalism and environmental history, and the implications for a framework to study humanity-in-nature.
2. Regression models implicitly assume zero measurement error, resulting in attenuated coefficients to the degree that error exists. SEM overcomes this limitation by associating an error term that represents random and nonrandom measurement error with each observed variable and assigning to endogenous latent variables a residual error term that reflects the effects of unmeasured variables in the model. As a result, path coefficients modeled in SEM are unbiased by error terms.

3. This time period is chosen based on the generally accepted view that "time lags of several decades" (Wackernagel et al. 2004:271) exist between the ecological changes and subsequent socioeconomic impacts. This value is derived by calculating a change score from 1971 to 2001 using the typical formula: $(T_2 - T_1)/T_1$. The data are then multiplied by negative one (*−1) to ease interpretation of results such that larger values indicate greater losses.
4. See Jorgenson and Clark (2012) for an exemplary illustration.
5. Though not presented, results available upon request.

References

Amin, Samir. 1974. *Accumulation on a World Scale*. New York: Monthly Review Press.

Auty, Richard M. 2001. "The Political Economy of Resource-Driven Growth." *European Economic Review* 45:839–946.

Baran, Paul A. 1957. *The Political Economy of Growth*. New York: Monthly Review Press.

Biel, Robert. 2006. "The Interplay between Social and Environmental Degradation in the Development of the International Political Economy." *Journal of World-Systems Research* 12(1):109–147.

Bollen, Kenneth. 1983. "World System Position, Dependency, and Democracy: The Cross-National Evidence." *American Sociological Review* 48:468–479.

———. 1989. *Structural Equations with Latent Variables*. New York: John Wiley.

Bonds, Eric and Liam Downey. 2012. "'Green' Technology and Ecologically Unequal Exchange: The Environmental and Social Consequences of Ecological Modernization in the World-System." *Journal of World Systems Research* 18(2):167–186.

Bornschier, Volker, Christopher Chase-Dunn, and Richard Rubinson. 1978. "Cross-National Evidence of the Effects of Foreign Investment and Aid on Economic Growth and Inequality: A Survey of Findings and a Reanalysis." *American Journal of Sociology* 84(3):651–683.

Bunker, Stephen G. 1985. *Underdeveloping the Amazon: Extraction, Unequal Exchange and the Failure of the Modern State*. Urbana: University of Illinois Press

Burkett, Paul and John Bellamy Foster. 2006. "Metabolism, Energy, and Entropy in Marx's Critique of Political Economy: Beyond the Podolinsky Myth." *Theory and Society* 35(1):109–156.

Byrne, Barbara N. 2009. *Structural Equation Modeling with AMOS*. New York: Routledge.

Catton, William R. 1980. *Overshoot: The Ecological Basis of Revolutionary Change*. Urbana, IL: University of Illinois Press.

Chase-Dunn, Christopher. 1975. "The Effects of International Economic Dependence on Development and Inequality: A Cross-National Study." *American Sociological Review* 40(6):720–738.

Collier, Paul and Anke Hoeffler. 2005. "Resource Rents, Governance, and Conflict." *Journal of Conflict Resolution* 49(4):625–633.

Crosby, Alfred W. 2004. *Ecological Imperialism: The Biological Expansion of Europe, 900–1900*. Cambridge, MA: Cambridge University Press.

Daly, Herman and John Cobb, Jr. 1989. *For the Common Good: Redirecting the Economy Toward Community, the Environment, and a Sustainable Future*. Boston, MA: Beacon Press.

De Soysa, Indra. 2000. "The Resource Curse: Are Civil Wars Driven by Rapacity or Paucity?" Pp. 113–136 in *Greed and Grievance: Economic Agendas in Civil Wars*, edited by Mats Berdal and David M. Malone. Boulder, CO: Lynne Rienner Publishers.

Diamond, Jared. 1997. *Guns, Germs and Steel: The Fates of Human Societies*. New York: W. W. Norton.

Dietz, Thomas and Andrew Jorgenson, editors. 2013. *Structural Human Ecology: New Essays in Risk, Energy, and Sustainability*. Pullman, WA: Washington State University Press.

Ding, Ning and Barry C. Field. 2005. "Natural Resource Abundance and Economic Growth." *Land Economics* 81(4):496–502.

Dixon, William J. and Terry Boswell. 1996. "Dependency, Disarticulation, and Denominator Effects: Another Look at Foreign Capital Penetration." *American Journal of Sociology* 102(2):543–562.

Dunning, Thad. 2005. "Resource Dependence, Economic Performance, and Political Stability." *Journal of Conflict Resolution* 49(4):451–482.

Foster, John Bellamy. 1999. "Marx's Theory of Metabolic Rift: Classical Foundations for Environmental Sociology." *American Journal of Sociology* 105(2):366–405.

Foster, John Bellamy, Brett Clark, and Richard York. 2010. *The Ecological Rift: Capitalism's War on the Earth*. New York: Monthly Review Press.

Foster, John Bellamy and Hannah Holleman. 2012. "Weber and the Environment: Classical Foundations for a Postexemptionalist Sociology." *American Journal of Sociology* 117(6):1625–1673.

———. 2014. "The Theory of Unequal Ecological Exchange: A Marx-Odum Dialectic." *Journal of Peasant Studies* 41(2):199–233.
Frank, Andre Gunder. 1978. *Dependent Accumulation and Underdevelopment*. New York: Monthly Review Press.
———. 2005. "Urban Location and Dissipation of Entropy." Accessed online at http://rrojasdatabank.info/agfrank/research.html#dissipation.
Frankel, Jeffrey A. 2010. "The Natural Resource Curse: A Survey." HKS Faculty Research Working Paper Series, RWP10-005, John F. Kennedy School of Government, Harvard University.
Frey, R. Scott. 1998. "The Hazardous Waste Stream in the World-System." Pp. 84–103 in *Space and Transport in the Modern World-System*, edited by Paul Ciccantell and Stephen Bunker. Westport, CT: Greenwood.
Georgescu-Roegen, Nicholas. 1971. *The Entropy Law and the Economic Process*. Cambridge: Harvard University Press.
Global Footprint Network. 2010. "Ecological Footprint and Biocapacity, 2006." Based on *National Footprint Accounts 2009*. Accessed online at www.footprintnetwork.org/atlas.
Grossmann, Henryk. 1929. *The Law of Accumulation*. London: Pluto Press.
Hornborg, Alf. 2001. *The Power of the Machine*. Walnut Creek, CA: Altamira Press.
———. 2014. "Ecological Economics, Marxism, and Technological Progress: Some Explorations of the Conceptual Foundations of Theories of Ecologically Unequal Exchange." *Ecological Economics* 105:11–18.
———. 2015. "Why Economics Needs to be Distinguished from Physics, and Why Economists Need to Talk to Physicists: A Response to Foster and Holleman." *Journal of Peasant Studies* 42(1):187–192.
Jorgenson, Andrew K. 2006. "Global Warming and the Neglected Greenhouse Gas: A Cross-National Study of Methane Emissions Intensity, 1995." *Social Forces* 84(3):1779–1798.
Jorgenson, Andrew K. and Brett Clark. 2012. "Are the Economy and the Environment Decoupling? A Comparative International Study, 1960–2005." *American Journal of Sociology* 118(1):1–44.
Jorgenson, Andrew K. and James Rice. 2007. "Uneven Ecological Exchange and Consumption-Based Environmental Impacts: A Cross-National Investigation." Pp. 273–288 in *Rethinking Environmental History: World-System History and Global Environmental Change*, edited by Alf Hornborg, J. R. McNeill, and Joan Martinez-Alier. Lanham, MD: Altamira Press.
Karl, Terry L. 1997. *The Paradox of Plenty: Oil Booms and Petro-States*. Berkeley, CA: University of California Press.

———. 2004. "Oil-led Development: Social, Political, and Economic Consequences." Pp. 661–672 in *Encyclopedia of Energy*, edited by Cutler J. Cleveland. New York: Elsevier.

Kentor, Jeffrey. 2001. "The Long Term Effects of Globalization on Income Inequality, Population Growth and Economic Development." *Social Problems* 48(4):435–455.

Lawrence, Kirk S. 2009. "The Thermodynamics of Unequal Exchange Energy Use, CO2 Emissions, and GDP in the World-System, 1975–2005." *International Journal of Comparative Sociology* 50(3–4):335–359.

Liu, Jianguo, Thomas Dietz, Stephen R. Carpenter, Marina Alberti, Carl Folke, Emilio Moran, Alice N. Pell et al. 2007. "Complexity of Coupled Human and Natural Systems." *Science* 317(5844):1513–1516.

Liu, Jianguo, Vanessa Hull, Mateus Batistella, Ruth DeFries, Thomas Dietz, Feng Fu, Thomas W. Hertel, et al. 2013. "Framing Sustainability in a Telecoupled World." *Ecology and Society* 18(2):26.

Marx, Karl. 1867/1977. *Capital, Volume 1*. New York: Vintage.

McKinney, Laura A. 2012. "Entropic Disorder: New Frontiers in Environmental Sociology." *Sociological Perspectives* 55(2):295–317.

———. 2014. "Foreign Direct Investment, Development, and Overshoot." *Social Science Research* 47:121–133.

McKinney, Laura and Kelly Austin. 2015. "Ecological Losses are Infecting Women: An Analysis of Female HIV Prevalence and Life Expectancy in Less-Developed Countries." *Social Problems* 62(4):529–549.

McNeill, John Robert. 2001. *Something New Under the Sun: An Environmental History of the Twentieth-Century World*. New York: W. W. Norton.

Moore, Jason W. 2000. "Environmental Crises and the Metabolic Rift in World-Historical Perspective." *Organization & Environment* 13(2):123–158.

———. 2015. *Capitalism in the Web of Life: Ecology and the Accumulation of Capital*. Brooklyn, NY: Verso Books.

Polterovich, Victor, Vladimir Popov, and Alexander Tonis. 2007. "Resource Abundance, Political Corruption, and Instability of Democracy." Working Paper No. 2007/73, New Economic School, Moscow, Russian Federation.

Prebisch, Raul. 1950. *The Economic Development of Latin America and Its Principal Problems*. New York: United Nations.

Prigogine, Ilya. 1967. *Introduction to Thermodynamics of Irreversible Processes*. New York: Interscience Publishers.

Rajan, Sudhir Chella. 2011. "Poor Little Rich Countries: Another Look at the 'Resource Curse'." *Environmental Politics* 20(5):617–632.

Rees, William E. 2004. "The Eco-Footprint of Agriculture: A Far-from-(Thermodynamic)-Equilibrium Interpretation." Pp. 87–110 in *NABC Report 16 Agricultural Biotechnology: Finding Common International Goals*, edited by Allan Eaglesham, Alan George Wildeman, and Ralph W. F. Hardy. Ithaca, NY: National Agricultural Biotechnology Council.

Rice, James. 2007. "Ecological Unequal Exchange: Consumption, Equity, and Unsustainable Structural Relationships within the Global Economy." *International Journal of Comparative Sociology* 48(1):43–72.

Robinson, James A., Ragnar Torvik, and Thierry Verdier. 2006. "Political Foundations of the Resource Curse." *Journal of Development Economics* 79(2):447–468.

Ross, Michael L. 1999. "The Political Economy of the Resource Curse." *World Politics* 51(2):297–322.

———. 2001. "Does Oil Hinder Democracy?" *World Politics* 53(3):325–361.

Rosser, Andrew. 2006. "The Political Economy of the Resource Curse: A Literature Survey." Working Paper 268. Brighton, UK: Institute of Development Studies.

Saad-Filho, Alfredo and John Weeks. 2013. "Curses, Diseases and Other Resource Confusions." *Third World Quarterly* 34(1):1–21.

Sachs, Jeffrey D. and Andrew M. Warner. 1995. "Natural Resource Abundance and Economic Growth." Working Paper No. 5398. Cambridge, MA: National Bureau of Research.

———. 2001. "The Curse of Natural Resources." *European Economic Review* 45(4–6):827–838.

Schultz, Jessica and Richard York. 2011. "Recognizing *Overshoot*: Succession of an Ecological Framework." *Human Ecology Review* 18(2):139–146.

Shandra, John M., Christopher Leckband, and Bruce London. 2009a. "Forest Exports and Ecologically Unequal Exchange: A Cross-National Analysis of Deforestation." *Organization & Environment* 22:293–310.

Shandra, John M., Christopher Leckband, Laura McKinney, and Bruce London. 2009b. "Ecologically Unequal Exchange, World Polity, and Biodiversity Loss: A Cross-National Analysis." *International Journal of Comparative Sociology* 50:285–310.

Shircliff, Eric J. and John M. Shandra. 2011. "Non-Governmental Organizations, Democracy, and HIV Prevalence: A Cross-National Analysis." *Sociological Inquiry* 81(2):143–173.

Singer, Hans W. 1950. "The Distribution of Gains Between Investing and Borrowing Countries." *The American Economic Review* XL:473–485.

Stiglitz, Joseph E. 2007. *Making Globalization Work*. New York: W. W. Norton.

Stijns, Jean-Philippe C. 2005. "Natural Resource Abundance and Economic Growth Revisited." *Resources Policy* 30:107–130.

United Nations Conference on Trade and Development (UNCTAD). 2014. "Foreign Direct Investment Statistics." Data for various years accessed online at unctad.org.

Wackernagel, Mathis, Chad Monfreda, Niels B. Schulz, Karl-Heinz Erb, Helmut Haberl, and Fridolin Krausmann. 2004. "Calculating National and Global Ecological Footprint Time-Series: Resolving Conceptual Challenges." *Land Use Policy* 21(3):271–278.

Wallerstein, Immanuel. 1974. *The Modern World-System I: Capitalist Agriculture and the Origins of the European World-Economy in the Sixteenth Century*. New York: Academic Press.

———. 2004. *World-Systems Analysis*. Durham, NC: Duke University Press.

World Bank. 2013. *World Development Indicators*. Data for various years accessed online at worldbank.org.

Wright, Gavin and Jesse Czelusta. 2004. "Why Economies Slow: The Myth of the Resource Curse." *Challenge* 47(2):6–38.

York, Richard, Eugene A. Rosa, and Thomas Dietz. 2003. "Footprints on the Earth: The Environmental Consequences of Modernity." *American Sociological Review* 68(2):279–300.

7

Mining Exports Flows, Repression, and Forest Loss: A Cross-National Test of Ecologically Unequal Exchange

Jamie M. Sommer, John M. Shandra, and Carolyn Coburn

According to the World Wildlife Fund (2016), approximately 50,000 square miles of forest are lost each year. This figure represents an area about the size of Greece and is equivalent to 48 football fields every minute (World Wildlife Fund 2016). However, forest loss is not equally distributed across the planet. Low- and middle-income nations have the highest rates of forest loss. From 1990 to 2010 forest area in high-income nations increased by 0.28 percent, while forest area in low- and middle-income nations decreased by 2.1 percent (World Bank 2015).

The theory of ecologically unequal exchange (EUE) offers a potential explanation for why this may be the case. According to this theory, rich nations are able to externalize their forest loss onto poor nations by importing natural resources and, as a result, contribute to higher rates of forest loss along with other environmental problems in poor nations (Bunker 1985). For instance, Jorgenson (2006) finds that weighted

J. M. Sommer (✉) • J. M. Shandra • C. Coburn
Department of Sociology, State University of New York at Stony Brook, Stony Brook, NY, USA
e-mail: jamie.sommer@stonybrook.edu; John.Shandra@stonybrook.edu; carolyn.coburn@stonybrook.edu

R. S. Frey et al. (eds.), *Ecologically Unequal Exchange*,
https://doi.org/10.1007/978-3-319-89740-0_7

exports from poorer to richer nations correspond with more forest loss. Jorgenson (2010) refines the previous study and finds that agriculture, forestry, and mining export flows are related to increased forest loss in poor nations. Jorgenson, Austin, and Dick (2009) use panel data and fixed effects regression models to confirm this finding. Further, Shandra, Leckband, and London (2009b) use sector-specific data and find that forestry export flows from poor to rich nations increase forest loss in poor nations. Finally, Austin (2010, 2012) uses data on specific commodities. She finds that the vertical flow of cattle exports (Austin 2010) and coffee exports (Austin 2012) from poor to rich nations corresponds with increased forest loss in poor nations. As we have outlined here, the cross-national research with its various refinements in measures and methods supports ideas drawn from the theory of EUE regarding the harmful impacts on forests.

The previous research served as the starting point for our study. However, we refine it in another way. We consider how mining exports sent from poor to rich nations affect forests in poor nations. This is a logical extension of existing work for several reasons. Bunker and Ciccantell (2005) describe in detail how iron-ore exports from Brazil destined for the United States led to increased forest loss in the Amazon. Further, data are available from the United Nations (2015) on mining export flows by sending and receiving country. In fact, Jorgenson (2010) and Jorgenson et al. (2009) include mining exports as part of their weighted export flow index in the primary sector.

At the same time, we seek to extend the cross-national research on EUE in another novel direction.[1] We consider how repressive nations facilitate EUE in the mining sector. We do so by drawing on ideas put forth in Evans's (1979) writing about the "triple alliance" of multinational capital, domestic capital, and the repressive government in Brazil during the 1970s. Following, Evans (1979) we argue that repressive nations create a "good business climate" for mining companies via economic incentives (e.g., tax holidays), regulatory concessions (e.g., environmental law exemptions and elimination of minimum wage), and imposed political stability (i.e., outlawing strikes, protests, and unions, and firing workers at will) so that mining export flows have a more adverse impact on forests in repressive than democratic nations. In developing this hypothesis, we draw on Downey, Bond, and Clark (2010), who provide detailed descriptions of how this process may play out in poor nations.

We use ordinary least squares regression to analyze data for a sample of 61 low-and middle-income nations from 1990 to 2010 on change in natural forest loss to test these hypotheses. We now turn to a discussion of the theoretical frame for this chapter and why we expect that mining export flows are associated with increased forest loss. We then consider how repressive nations may facilitate unequal exchange in the mining sector and the implications for forests. We conclude by describing the variables, data and method, and findings.

Ecologically Unequal Exchange Theory, Mining, and Forest Loss

In recent years, a large volume of theory and empirical research describes how the structure of international trade contributes to various forms of environmental degradation. The theory of EUE is one of the main orientations in this burgeoning area of inquiry. Given its focus on trade relationships between rich and poor nations, this perspective has its origins in the dependency and world-systems traditions (Amin 1976). However, it was Bunker (1985), who first described how exports sent from poor to rich nations tend to affect adversely the natural environment of poor nations, in his book titled *Underdeveloping the Amazon: Extraction, Unequal Exchange, and the Failure of the Modern State.*

Why may this be the case? In general, wealthy nations tend to be advantageously situated within the global economy and are more likely to secure favorable terms of trade (Bunker 1985; Hornborg 2003). This advantage is because the prices of exports from poor nations, largely natural resources and agricultural products, consistently fall relative to the prices of items exported by wealthy nations that largely include manufactured goods. This decline in terms of trade is the result of several factors that drive down the price of natural resources and agricultural goods. These factors may include large numbers of poorer nations producing similar products, subsidies by governments in rich nations producing similar products, and an abundant supply of cheap labor to produce the goods in poor nations. The World Bank and International Monetary Fund facilitate such exchange via their structural adjustment lending. These loans require borrowing nations to increase exports of natural resources by devaluing currency by liberalizing trade by offering various economic incentives (e.g., tax holidays) and regulatory concessions (e.g., environmental law exemptions) to foreign investors (Shandra, Shircliff, and London 2011). Unequal exchange is also exacerbated by the ability or threat of a multinational corporation to relocate their operations to a nation offering a more favorable investment climate than the nation in which its current operations are located (Clapp and Dauvergne 2005).

In the end, it takes increasingly larger levels of agricultural and natural resource exports to buy imports from rich nations (Muradian and Martinez-Alier 2001). A poor nation can be very successful at exporting more natural resources, but, in return for the sale of those natural resources, it gets fewer, not more, imports (Giljum and Eisenmenger 2004). This often translates into extensive environmental degradation within the boundaries of poor nations (e.g., forest loss, water pollution, and air pollution) as they expand export production just to maintain current levels of imports (Roberts and Parks 2007).

The empirical analyses of EUE theory have become quite popular among cross-national scholars. In fact, we describe the evolution of this work over the past ten years or so as it pertains to forest loss in the introduction of the chapter and note that it serves as the starting point for our study. However, we refine it yet again by considering if mining exports sent from poor to rich nations increase forest loss in poor nations.

Returning to Bunker and Ciccantell's (2005) research, Brazil with financing from the World Bank built an iron-ore mine, smelters, and a railway connecting the mine to port to load the smelted iron on to ships destined for the United States. It also included the creation of settlements in an effort to "develop" this region (Bryant and Bailey 1997). Over 1500 square miles of forests were cleared annually to make way for the mine, smelter, and related settlements along with meeting the demand for charcoal to power the smelters (Rich 1994).

From this brief review of the theory of EUE, we hypothesize that higher levels of mining exports flow from poor to rich nations should be associated with higher rates of forest loss in poor nations.

Possible Interaction Effects

What role may governments play in the EUE of exports? What are the implications for forests? In some of the earliest writings on the topic, Evans (1979) described the existence of a "triple alliance" of multinational capital, local capital, and the repressive government in Brazil. This alliance rested upon the government creating a "good business climate" by offering investors a variety of economic incentives and regulatory concessions.

The most notable economic incentives entail "tax holidays" or waiving corporate taxes and tariffs for a period of time, while regulatory concessions entail exemptions to environmental regulations, labor flexibility measures, and appropriation of land (London and Ross 1995). At its most extreme, repressive nations seek to control their populations using violence by the military (Downey et al. 2010). In the end, leaders of repressive nations are able to finance industrialization, exert further control over remote territory, and increase their personal wealth (Evans 1979; O'Connor 1998). Multinational corporations operate in nations that help them keep costs low (Evans 1979). Nevertheless, it often results in increased forest loss within a poor nation.

Downey et al. (2010) describe how repressive governments of several poor nations use a variety of regulatory concessions in order to facilitate mining exports to rich nations. For instance, a Rio Tinto copper mine in Papua New Guinea led to the loss of thousands of acres of forest, billons of tons of mine waste dumped into local rivers, and, ultimately, health issues among the country's population (Downey et al. 2010). However, the government suppressed protests directed at Rio Tinto over these issues by firing miners, replacing them with foreign workers, and forcibly relocating local workers to elsewhere in the country (Downey et al. 2010).

In Gabon, the government has restricted reporting on mining and movement into areas where manganese is mined for export to the United

States (Downey et al. 2010). Nevertheless, it is causing substantial forest loss in areas surrounding the mine along with water and air pollution (Downey et al. 2010). A similar situation has played out around the exporting of diamonds from Zimbabwe, coal from Colombia, and nickel from Indonesia (Downey et al. 2010).

It is important to note that repressive nations also provide mining companies with economic incentives that facilitate the EUE and, ultimately, forest loss. For instance, companies mining in the Democratic Republic of the Congo pay substantially less corporate income taxes than companies in other sectors along with lower tariffs on equipment imported to the country (Butler 2012). It is estimated that the government lost $1.36 billion in revenue as a result of these tax breaks. At the same time, forest loss increased around mines (Butler 2012). Similarly, in Mongolia, the government encourages mining of gold and copper by providing small- and medium-sized companies with tax breaks in an effort to keep exports of these minerals high (Snow 2015). However, it has resulted in forest loss in the country along with a host of other environmental problems (Snow 2015).

We draw on Evans's (1979) writing on the triple alliance in Brazil and examples from Downey et al. (2010) to demonstrate that repressive governments put into place a "good business climate" that facilitates the EUE of mining exports. This leads us to the hypothesis that mining exports from poor to rich nations increase forest loss more in repressive than democratic nations. We now turn to a discussion of the data and method we used to test our hypotheses.

Data and Method

Dependent Variable

The dependent variable is the average annual percentage change in natural forest area from 2000 to 2010—the most recent data available on forest loss (Food and Agriculture Organization 2015). Please note that deforestation is signified by a positive value for interpretation purposes. We created this value by multiplying the original measure by a value of negative one. This measure includes land greater than half a hectare in size with trees higher than 5 meters and a canopy cover of more than 10 percent. A natural forest consists only of native forest species with the possible exception of small areas of natural regeneration or assisted natural regeneration. This measure excludes forest plantations, which are areas established through planting or seeding (Food and Agriculture Organization 2015). A forest plantation often involves relative homogeneity in the types of species grown for commercial purposes (World Resources Institute 2005). We used natural forest area data since we are interested in the potential effects of mining export flows on forests that are not being intensively managed for commercial production (e.g., forest plantations). In Table 7.1, we provide descriptive statistics and a bivariate correlation matrix for all the variables used in the analysis.

Independent Variables

Mining export flows. We included the flows of mining exports from rich to poor nations to test EUE theory. In particular, this variable measures a nation's mining exports sent to Organization for Economic Cooperation and Development (OECD) nations as a percentage of a nation's total mining exports. The data are from the United Nations (2008) *Commodity Trade Statistics Database.* This database reports import and export statistics in US dollars for nations by commodity and trading partner. We use the first revision of the Standard International Trade Classification to identify mining sector exports. These include data on ferrous and nonferrous minerals. For some poor nations, there is incomplete information on mining exports. To deal with this potential limitation, we follow Moore, Teixeira, and Schiell's (2006) practice of using import data from trading partners to reconstruct missing export data. According to EUE theory, we hypothesize that higher levels of mining exports sent from poor to rich nations are associated with higher rates of forest loss in poor nations.

Repression. We use data from the Polity IV dataset to measure democracy (Marshall and Jaggers 2006). The democracy variable ranges from negative ten (autocracy) to ten (democracy). We multiple this variable by negative

Table 7.1 Descriptive statistics and bivariate correlation matrix for deforestation analysis ($N = 61$)

	Mean	Standard deviation	(1)	(2)	(3)	(4)	(5)	(6)	(7)	(8)	(9)	(10)	(11)	(12)	(13)	(14)	(15)
(1) Deforestation, 1990–2010	0.123	0.168	1.000														
(2) Mining export flows	0.701	0.282	0.014	1.000													
(3) Repression, 1990	−0.164	7.167	0.306	−0.150	1.000												
(4) Non-governmental organizations, 1990	3.869	3.389	−0.205	0.037	−0.449	1.000											
(5) Environmental ministry, 1990	0.426	0.499	−0.077	0.038	0.067	−0.026	1.000										
(6) Forestry exports, 1990	0.002	0.006	−0.097	0.023	0.012	0.256	0.193	1.000									
(7) Gross domestic product, 1990	7.564	0.808	−0.353	0.296	−0.708	0.485	0.056	0.176	1.000								
(8) Economic growth, 1980–1990	2.425	0.512	0.327	−0.063	0.082	0.026	−0.065	0.122	−0.175	1.000							
(9) Total population growth, 1980–1990	0.262	0.107	0.459	0.003	0.537	−0.413	0.025	−0.072	−0.517	0.266	1.000						
(10) Non-dependent population growth, 1980–1990	0.289	0.107	0.375	0.025	0.346	−0.155	0.006	0.062	−0.296	0.371	0.884	1.000					
(11) Rural population growth, 1980–1990	0.145	0.151	0.423	−0.216	0.511	−0.425	0.001	−0.159	−0.656	0.140	0.752	0.646	1.000				
(12) Urban population growth, 1980–1990	0.513	0.262	0.346	−0.067	0.598	−0.387	0.063	0.017	−0.604	0.341	0.652	0.547	0.308	1.000			
(13) Population density, 1990	89.507	136.133	−0.100	−0.354	−0.165	0.130	0.029	−0.053	0.040	0.213	−0.225	−0.114	−0.004	−0.156	1.000		
(14) Natural forest stocks, 1990	8.808	1.921	−0.043	0.096	0.158	0.288	−0.056	0.357	−0.117	−0.010	0.200	0.248	0.026	0.148	−0.472	1.000	
(15) Tropical climate, 1990	76.885	39.286	0.226	−0.009	0.351	−0.300	0.055	−0.031	−0.338	0.335	0.370	0.273	0.368	0.226	0.108	−0.057	1.000

one so that high scores correspond with a high level of repression. According to Li and Reuveny (2006), higher levels of democracy decrease forest loss because democratic nations have high levels of political activism. This is the case because democracies guarantee certain rights to their citizens, including freedoms of speech, press, and assembly. The leaders of a democracy must be responsive to such activism because of electoral accountability (Midlarsky 1998). Further, greater freedom of the press and assembly leads to a wider diffusion of information, which raises public awareness especially around environmental issues (Ehrhardt-Martinez, Crenshaw, and Jenkins 2002). We expect that higher levels of repression correspond with higher rate of forest loss.

Non-governmental organizations. We included the number of international non-governmental organizations working on "environmental" and "animal rights" issues per capita in a nation for 1990. The data were collected by Smith and Wiest (2005) from the *Yearbook of International Associations.* It is important to note that the data excludes labor unions, institutes, and foundations. Schofer and Hironaka (2005) find that higher levels of non-governmental organizations are associated with lower rates of deforestation. This may be the case because non-governmental organizations finance local conservation projects, support social movement activity around environmental issues, shape the language of environmental agreements, and write codes of conduct (Shandra 2007).

Environmental ministry. We also included a dummy variable that designates whether or not a nation has an environmental ministry. We coded nations that have an environmental ministry in 1990 with a value of one. All other nations serve as the reference category and are coded with a value of zero. The data were obtained from Frank (1999). We hypothesized that nations with an environmental ministry should be associated with lower rates of deforestation than nations without an environmental ministry since environmental ministries tend to implement programs that can reduce deforestation. Programs may include demarcating protected areas, monitoring of forests for illegal logging, and monitoring compliance with forestry regulations (Hurst 1990).

Gross domestic product. We employed a measure of gross domestic product per capita for 1990 to control for the level of development. We logged this variable to correct for its skewed distribution. Burns, Kick, and Davis (2003) have reported that higher levels of economic development are associated with lower rates of deforestation and argue that this finding is a result of wealthier nations "externalizing" their environmental costs by importing natural resources from poorer nations.

Economic growth. We included the average annual economic growth rate from 1990 to 2000. This measure was taken from the World Bank (2005). We logged the variable because it is skewed. It is generally thought that economic growth should be associated with higher rates of deforestation. This relationship is positive because there are large amounts of capital available for investment in activities that accelerate forest loss during periods of economic expansion (Rudel 1989). Jorgenson (2006) has reported that economic growth is associated with increased forest loss.

Forestry exports. We included the value of forestry exports as a percentage of total exports. This measure includes exports of chips, fiberboard, industrial round wood, paper, plywood, particle board, sawnwood, veneer sheets, and wood pulp (United Nations 2016). The forestry data are for 1990 and were obtained from the United Nations (2016). The data on total exports come from the World Bank (2005). We included this variable because it has been reported to be associated with increased forest loss (Allen and Barnes 1985).

Total population growth. Demographic factors have often been found to be associated with forest loss. Therefore, we included the percentage change in total population growth from 1980 to 1990 in the analysis. Data were taken from the World Bank (2005). The general argument is that "geometric" growth in population outstrips "arithmetic" growth in the means of subsistence, leading to "carrying capacity" problems and ensuing environmental problems (e.g., forest loss) (Rudel 1989). Thus, we expected higher rates of population growth correspond with higher rates of forest loss. Rudel (1989) has reported support for this hypothesis.

Non-dependent population growth. York, Rosa, and Dietz (2003) have argued that it is important to "decompose" demographic factors in cross-national studies. That is, researchers should examine not just overall growth rates per se but also the impact of population growth in different contexts. York et al. (2003) found that the higher levels of a nation's non-dependent population (i.e., population aged 15–64 years) increase its ecological footprint. Thus, we decomposed the total population growth in our analysis and included the percentage change of a nation's non-dependent population from 1980 to 1990 in the regression models. Data were taken from the World Bank (2003). We expected that non-dependent population growth is correlated with higher rates of forest loss.

Rural and urban population growth. Jorgenson and Burns (2007) decomposed population growth differently and found that higher rates of rural population growth are associated with increased forest loss, while higher rates of urban population growth are associated with lower rates of deforestation. They argued that expanding urban centers often creates economic opportunities other than agricultural ones, which leads to increased rural to urban migration. Thus, we included the percentage change in rural and urban populations between 1980 and 1990 in the models.

Population density. According to Ehrhardt-Martinez et al. (2002), it is important to measure how a population is distributed. In this regard, we also controlled for population density or total population of a nation divided by land in square kilometers for 1990 (World Bank 2003). We expected that higher levels of population density are associated with lower rates of forest loss because highly concentrated populations should take pressure off forests (Ehrhardt-Martinez et al. 2002).

Tropical climate. We included the percentage of land in a nation that is defined as tropical. The data were taken from the World Resources Institute (2005), which defines a tropical climate as land area that has a mean monthly temperature that exceeds 18°C. We logged this variable because it is skewed. We hypothesized that tropical nations have higher rates of deforestation because these nations tend to have more valuable tree species that are in demand on the world market. Rudel and Roper (1997) found support for this hypothesis.

Natural forest stocks. It is necessary to include a measure that controls for the potentially biasing effects of relative abundance or scarcity of forest resources (Rudel 1989). Therefore, we included the percentage of a nations' land area covered by forests or the amount of natural forest stocks in hectares when examining change in natural forest area. The data for both measures were taken from the Food and Agriculture Organization (2010).

Method of Analysis

We used ordinary least squares regression to analyze data for the 61 low- and middle-income nations included in the sample (Stata version 13).[2] This is the most common method employed to analyze the determinants of forest loss (e.g., Austin 2010).

This model estimated is denoted by the following formula:

$$y_i = a + b_1 X_1 + b_2 X_2 + \cdots + b_k X_k + e_i$$

where

y_i = dependent variable for each country
a = the constant
B_1 to B_k = unstandardized coefficients for each independent variable
X_k = independent variables for each country
e_i = error term for each county

Since we used ordinary least squares to analyze the data, we first examined whether regression assumptions were violated. We examined the following: multicollinearity, linearity,[3] outliers, influential cases,[4] and heteroscedasticity.

First, we calculated mean and highest variance inflation factor scores for each model. Multicollinearity is not a problem if mean and highest variance inflation factor scores are less than 2.5 (York et al. 2003). In our models, the average variance inflation factor scores did not exceed this threshold. However, they did exceed this threshold when interaction terms were included in estimates. This is not surprising. However, we reran the models centering both variables, which greatly reduce the variance inflation factor scores in the interactive models. Therefore, multicollinearity appears to be present but not problematic in the models (Allison 1999).

Second, we examined scatterplots of each independent variable against the dependent variable, and calculated skewness statistics for each variable to determine if there are any problems with linearity (Allison 1999).

We transformed variables when appropriate with the natural log (Tabachnick and Fidell 2013). We noted any transformation in the variable description section above.

Third, we determined if unusual data points affected the results. We began by calculating standardized residuals to determine if outliers were a problem. There were no nations with standardized residuals greater than an absolute value of three (Tabachnick and Fidell 2013). We also calculated Cook's distance statistics to ensure there are no influential cases. There was no country with a distance statistic that exceeded one, indicating an absence of influential cases (Tabachnick and Fidell 2013).

Fourth, we calculated Breush-Pagan statistics for each model to determine if heteroscedasticity was a problem. The null hypothesis for this chi-square test is that the error variances are homoscedastic or equally distributed. The coefficients for these chi-square statistics reached a level of statistical significance in most models, indicating potential problems with heteroscedasticity with ordinary least squares estimates (Tabachnick and Fidell 2013). We present robust standard errors to help deal with this issue.

Findings

Ordinary least squares regression estimates with robust standard errors of forest loss are reported in Table 7.2. The first number presented is the unstandardized coefficient, the second number is the standardized coefficient, and the third number in parentheses is the robust standard error. We report one-tailed hypothesis tests because of the directional nature of our hypotheses.

In every equation, we include mining export flows,[5] democracy, non-governmental organizations, environmental ministry dummy variable, forestry exports,[6] gross domestic product per capita,[7] economic growth,[8] population density, forest stocks, and the tropical climate dummy variable.[9] In Equation (7.2.1), we examine the impact of total population growth. In Equation (7.2.2), we examine the impact of non-dependent population growth. In Equation (7.2.3), we consider the impact of rural and urban population growth.

Table 7.2 Linear models of mining export flows on forest loss (1990–2010)

	Equation (7.2.1)	Equation (7.2.2)	Equation (7.2.3)
Independent variables			
Mining export flows	0.035	0.036	0.050
	0.058	0.060	0.084
	(0.071)	(0.074)	(0.074)
Repression, 1990	0.001	0.002	0.002
	0.051	0.078	0.072
	(0.005)	(0.005)	(0.005)
Non-governmental organizations, 1990	0.008	0.005	0.007
	0.160	0.108	0.145
	(0.007)	(0.008)	(0.008)
Environmental ministry, 1990	−0.023	−0.020	−0.025
	−0.070	−0.059	−0.075
	(0.041)	(0.041)	(0.041)
Forestry exports, 1990	0.071	−0.566	−0.508
	0.003	−0.021	−0.019
	(3.218)	(3.024)	(3.022)
Gross domestic product, 1990	−0.046	−0.056	−0.007
	−0.222	−0.271	−0.036
	(0.035)	(0.035)	(0.033)
Economic growth, 1980–1990	0.074*	0.073*	0.084*
	0.229	0.224	0.257
	(0.039)	(0.041)	(0.046)
Total population growth, 1980–1990	0.540*		
	0.345		
	(0.304)		
Non-dependent population growth, 1980–1990		0.380	
		0.243	
		(0.243)	
Rural population growth, 1980–1990			0.404*
			0.366
			(0.187)
Urban population growth, 1980–1990			0.097
			0.152
			(0.137)
Population density, 1990	−0.001*	−0.001*	−0.001*
	−0.188	−0.223	−0.216
	(0.001)	(0.001)	(0.001)
Natural forest stocks, 1990	−0.025	−0.025	−0.021
	−0.290	−0.283	−0.239
	(0.017)	(0.017)	(0.017)

(*continued*)

Table 7.2 (continued)

	Equation (7.2.1)	Equation (7.2.2)	Equation (7.2.3)
Tropical climate, 1990	−0.001 −0.015 (0.001)	0.001 0.009 (0.001)	−0.001 −0.239 (0.001)
Constant	0.352 (0.325)	0.464 (0.332)	0.025 (0.287)
R-squared	0.328	0.306	0.335
Number of cases	61	61	61

Notes:
* indicates $p < 0.05$ for a one-tailed test
The first number is the unstandardized coefficient, the second is the standardized coefficient, and the robust standard error is in parenthesis

Let us begin by talking about mining export flows from poor to rich nations. We find little support for EUE theory that mining exports sent from poor to rich nations correspond with forest loss in poor nations. The coefficients from this export flow measure fail to reach levels of statistical significance in both models. This is somewhat surprising given the case study evidence, and the utility of the theory being applied to the flow of other types of exports.

We do, however, find several other factors related to forest loss. First, we find that economic growth is associated with increased forest loss. The coefficients for this variable are positive and significant in every model. These results support the idea that large capital investments in activities that accelerate forest loss during periods of economic expansion (Rudel 1989). Second, we find that total population growth is associated with increased forest loss as well. In Equation (7.2.1), the coefficient for this measure is positive and statistically significant. Third, the coefficient for rural population growth is positive and significant in Equation (7.2.3). This finding supports ideas that growing rural populations clear forests to grow crops or make way for other forms of economic activity (Jorgenson and Burns 2007). Fourth, we find that higher levels of population density correspond with lower levels of forest loss. The coefficients for this variable are negative and significant in every model. Thus, nations with more densely populated areas tend to have less forest loss (Ehrhardt-Martinez et al. 2002).

There are some non-significant findings. First, several economic factors are not related to forest loss. These include gross domestic product per capita and forestry exports. Second, there are political factors that fail to explain significant variations in forest loss. The coefficients for non-governmental organizations, democracy, and environmental ministry fail to reach levels of statistical significance. Third, we do not find that forest stocks or tropical land area are related to forest loss. The coefficients are not significantly different from zero in any of the models in Table 7.2.

The non-significant finding regarding the mining export flow variable is somewhat surprising given substantial theory and empirical evidence in the area of EUE. Up to this point, we have only considered the additive impact of mining export flows on forest loss. However, as we noted above it may be necessary to examine an *interactive* relationship between repression and mining flows to fully understand how these factors impact forest loss. We argue that this is because repressive nations create a "good business climate" for corporations that facilitate the unequal exchange of mining exports by providing imposed political stability, economic incentives, and regulatory concessions. Toward this end, we expect that mining exports increase forest loss more at higher levels than at lower levels of repression.

In Table 7.3, we test this hypothesis by including an interaction term between mining export flows and repression in the models. We construct the interaction terms by centering the moderator variables (i.e., repression/democracy) around its mean. We then multiply the centered version of the moderator variable by the focal variable (i.e., mining export flows) (Jaccard 2001). We provide estimates of the effect mining export flows at low (i.e., −10), medium (i.e., zero), and high (i.e., 10) levels of repression. By centering the moderator variable (i.e., repression), the effect of mining export flows at medium levels of repression is equal to the unstandardized coefficient for the mining export flow variable (Jaccard 2001). It is important to note that Table 7.3 has the same organization as Table 7.2.

What do these model specifications reveal? We find that mining export flows increase forest loss more in repressive than democratic nations. First, the coefficients for the interaction terms are statistically significant in all three models. Second, when we calculate the impact of mining

Table 7.3 Interactive models of mining flows on forest loss (1990–2010)

	Equation (7.3.1)	Equation (7.3.2)	Equation (7.3.3)
Independent variables			
Mining export flows	−0.012	−0.016	0.005
	−0.021	−0.027	0.009
	(0.077)	(0.080)	(0.080)
Repression, 1990	−0.013	−0.013	−0.011
	−0.536	−0.567	−0.481
	(0.008)	(0.008)	(0.008)
Non-governmental organizations, 1990	0.009	0.006	0.008
	0.172	0.122	0.157
	(0.007)	(0.007)	(0.007)
Environmental ministry, 1990	−0.014	−0.010	−0.017
	−0.042	−0.029	−0.050
	(0.042)	(0.042)	(0.041)
Forestry exports, 1990	−1.084	−1.803	−1.636
	−0.040	−0.066	−0.060
	(3.472)	(3.245)	(3.235)
Gross domestic product, 1990	−0.044	−0.054	−0.005
	−0.211	−0.261	−0.026
	(0.035)	(0.034)	(0.033)
Economic growth, 1980–1990	0.074*	0.070	0.082
	0.225	0.213	0.250
	(0.042)	(0.044)	(0.049)
Total population growth, 1980–1990	0.550*		
	0.351		
	(0.291)		
Non-dependent population growth, 1980–1990		0.421*	
		0.269	
		(0.229)	
Rural population growth, 1980–1990			0.394*
			0.355
			(0.185)
Urban population growth, 1980–1990			0.109
			0.170
			(0.134)
Population density, 1990	−0.001*	−0.001*	−0.001*
	−0.198	−0.235	−0.225
	(0.001)	(0.001)	(0.001)
Natural forest stocks, 1990	−0.024	−0.024	−0.019
	−0.275	−0.272	−0.223
	(0.016)	(0.016)	(0.016)

(*continued*)

Table 7.3 (continued)

	Equation (7.3.1)	Equation (7.3.2)	Equation (7.3.3)
Tropical land area, 1990	0.001	0.001	0.001
	0.027	0.054	0.035
	(0.325)	(0.001)	(0.001)
Interaction terms and calculated effects			
Mining export flows × Repression	0.018*	0.020*	0.017*
	0.602	0.658	0.564
	(0.009)	(0.010)	(0.010)
Calculated mining export flows effect at high repression	0.168	0.184	0.175
Calculated mining export flows effect at low repression	−0.201	−0.216	−0.165
Constant	0.345	0.461	0.019
	(0.345)	(0.321)	(0.283)
R-squared	0.363	0.348	0.366
Number of cases	61	61	61

Notes:
* indicates $p < 0.05$ for a one-tailed test
The first number is the unstandardized coefficient, the second is the standardized coefficient, and the robust standard error is in parenthesis

export flows at different levels of repression, we find that they more adversely impact forests at higher rather than lower levels of repression. In Equation (7.3.1), we find that that the calculated effect of mining exports flows on forest at high levels of repression is equal to 0.168, and equal to −0.201 at low levels of repression. In Equation (7.3.2), the effect of mining export flows on forest loss at high levels of repression is equal to 0.184 while it is equal to −0.216 at low levels of repression. In Equation (7.2.3), we find a similar pattern with the interaction term and related calculations. Finally, it is important to note that the findings from Table 7.2 remain stable across the new model specifications.

Discussion and Conclusion

We began by reviewing the cross-national research that finds that EUE of exports adversely affects forests in poor nations. This work serves as the starting point for our study. However, we refine it in a slightly different way than in previous research. We consider how the flow of mining exports from poor to rich nations affects forests in poor nations. In doing so, we initially find little support for the theory using this refinement. This is somewhat surprising given the theory and empirical work on the topic.

We extend the thinking on EUE by integrating ideas from Evans's (1979) work on the triple alliance and examples from Downey et al. (2010). From these works, we tested the hypothesis that mining export flows are associated with more forest loss at higher levels of repression than at lower levels of repression. When including interaction terms between these variables in the models, we found substantial support for this line of reasoning. The coefficients for the interaction terms are statistically significant in every model and calculations reveal that mining export flows increase forest loss at higher rather than lower levels of repression.

There are some theoretical implications of the findings. In a cross-national study on urbanization, London (1987:55) writes "international and intranational dynamics are so interpenetrating in the modern world system that any analysis that does not consider such interactions is seriously deficient and offers at best a partial explanation." We agree with London and go one step further by arguing that there must be a willingness by sociologists to self-consciously rid themselves of the sort of "theoretical blinders" that lead to the categorical analysis of how international factors (e.g., mining export flows) and intranational factors (e.g., repression) affect the natural environment. This is the case because these "ostensibly antithetical factors are related to each other in a specifiable and meaningful manner" (London and Smith 1988:44). It is only by considering possible interactions among internal and external characteristics of a nation that we will arrive at the most comprehensive understanding of the factors shaping natural environment and be able to refine theory accordingly.

There are some policy implications of the research reported here. The results suggest that democracy mitigates the harmful impact of mining exports flows on forests. There are certain policies that governments can promote that may lessen forest loss caused by exports, multinational corporations, and structural adjustment. Such policies include (1) popular participation in decision making; (2) public access to information; (3) recognition of labor unions; and (4) guarantees of free press, speech, and assembly (Rich 1994).

A nice illustration of how such laws may reduce forest loss comes from Brazil. In 1986, Brazil passed an environmental impact assessment law that not only called for their integration into development projects but also required that the assessments be written in publicly understandable language and public hearings to be held to discuss the results (Rich 1994). The law was ignored by the government for many years despite its passage. However, approximately 500 people assembled for the first time in 1990 to discuss a cattle ranching proposal (Rich 1994). The cattle ranchers applied for a permit to convert a large amount of forest to pasture. The National Council of Rubber Tappers along with the help of international non-governmental organizations drafted a detailed refutation of the environmental impact report for the plan (Barbosa 2001). Following the meeting, the state environmental agency ruled against the forest loss proposal (Rich 1994).

The strengthening of such laws in poor nations needs to be coupled with policy that seeks to reduce consumption of mineral exports in rich nations. The Climate Alliance of European Cities serves as an informative example in regard to reducing forestry exports from poor nations. This non-governmental organization convinced 200 municipalities in Germany and Austria to ban the use of tropical timber in government financed projects and to reduce greenhouse gas emissions in the municipalities below national levels for 1990 (Rich 1994). The Climate Alliance of European Cities also provides funding to the Instituto de Pre-Historia, Antropologia, e Ecologia, a Brazilian non-governmental organization, to rehabilitate logged forests by replanting local tree species, to support farmers in raising tree crops, to demarcate extractive reserves, and to monitor protected areas for illegal forest extraction (Rich 1994). A similar campaign could be applied to governments and companies that purchase goods and services from mining companies that operate in repressive nations. This may be even more effective if combined with a consumer boycott against mining companies or companies that purchase minerals exports from operations taking place in repressive nations (Smith and Wiest 2005).

Notes

1. We limit our discussion to the cross-national literature on forest loss because it is the most relevant and considerations of space. However, we do acknowledge that there is a larger cross-national literature on EUE that includes the analysis of ecological footprints (e.g., Jorgenson 2005), carbon dioxide emissions (e.g., Jorgenson 2012), industrial water pollution (e.g., Shandra, Shor, and London 2009c), and biodiversity loss (e.g., Shandra et al. 2009a). Austin (2012) demonstrates that EUE adversely impacts not only forests but also educational attainment and malnutrition.
2. The following 61 nations are included in the analysis. They are Algeria, Angola, Argentina, Bangladesh, Benin, Bolivia, Brazil, Bulgaria, Burkina Faso, Central African Rep, Chad, Chile, China, Colombia, Congo, Cote d'Ivoire, El Salvador, Gabon, Gambia, Ghana, Guatemala, Guinea, Guinea-Bissau, Honduras, Hungary, India, Indonesia, Jamaica, Kenya, Madagascar, Malawi, Malaysia, Mali, Mauritius, Mexico, Mauritania, Mongolia, Mozambique, Nepal, Nicaragua, Nigeria, Pakistan, Panama, Papua New Guinea, Paraguay, Peru, Philippines, Romania, Rwanda, Senegal, Sierra Leone, Sri Lanka, Tanzania, Thailand, Togo, Trinidad, Turkey, Uganda, Uruguay, Zambia, and Zimbabwe.
3. We deal with issues of linearity by taking the natural log of skewed variables. We acknowledge that this complicates the interpretation of the effects of variables. One way to deal with this issue is to use an elasticity model where all variables are logged with the exception of dummy variables (York et al. 2003). We reran the model using an elasticity model and results are similar. However, we do see a marked drop in the R-square values, indicating potential problems with model fit. This is not surprising as most variables approximate linearity prior to transformation and taking the natural log yields a relationship that does not approximate linearity. To further increase the reliability of the results, however, we also run the models with no variables transformed and keeping the original metrics for everything. This (again not ideal since some variables are not linear) yields substantively similar findings to the model presented but with substantially lower R-square values.
4. We also ran the analysis using a robust regression model that uses iteratively reweighted least squares with Huber and biweight functions tuned for 95 percent Gaussian efficiency to ensure there are no problems with

outliers (Dietz, Frey, and Kalof 1987). Put simply, this model assigns a weight to each observation used in the analysis with higher weights given to observations with smaller residuals and lower weights given to observations with higher residuals (Dietz et al. 1987). The results of the robust regression are similar to the results obtained by ordinary least squares. This enhances the reliability of the findings. We report the ordinary least squares results because they are more easily interpreted.

5. We use a weighted mining export flow variable. It is possible that this measure may be capturing the same underlying structural relationships as total export flows. However, we do not think this is the case. First, we find a fairly low bivariate correlation between mining export flows and export flow measures in other sectors. For instance, the bivariate correlation between mining and agricultural export flows equals 0.104 in our sample. The bivariate correlation with forestry export flows is equal to 0.130 in the sample. The bivariate correlation with total exports flows from poor to rich nations is equal to 0.150 for the sample. At the same time, there is a theoretical rationale for considering only mining exports in a study of forest loss. We review the reasons why EUE theory expects mining export flows to increase forest loss. However, there is research suggesting that oil, gas, and mineral exports may have a negligible impact on forest loss (Rudel 2013). Rudel (2013) argues that the discovery of mineral, oil, or natural gas deposits in sub-Saharan Africa triggers economic booms in extractive sectors, which slows growth in the agricultural sector. This is because extractive sector booms create a demand for labor, which makes it more expensive in a country. With workers seeking higher wages in mines or oil fields, labor shortages emerge especially in agriculture where wages tend to be lower. As a result, fields lay fallow (Rudel 2013). Further, oil, gas, and mineral exports tend to drive up the price of a country's currency, which makes its agricultural exports more expensive relative to a country with a devalued currency exporting similar crops (Rudel 2013). In both instances, there is less pressure to clear forests to expand agriculture (Rudel 2013).
6. We consider how other aspects of a country's exports may affect forest loss. These include total mining exports and total exports. The coefficients for these variables fail to reach a level of statistical significance.
7. We examine how other measures of a country's macroeconomic conditions affect forest loss. These include international trade, balance of payments, currency reserves, and exchange rates. These data can be obtained

from the World Bank (2015). The coefficients for each measure failed to reach a level of statistical significance, but the other results are similar to the findings in Table 7.2.

8. We include economic activity value added from agriculture, forestry, and mining as a percent of gross domestic product to account for a country's domestic economy structure. We would expect this measure to be associated with higher rates of deforestation because these types of economic activities are presumed to put greater pressure on forests than other types of economic activities (e.g., services and manufacturing). This variable does not explain any significant variation in forest loss.
9. We classify forestry statistics as being highly reliable if they are based on remote sensing survey or current national field sampling estimates (Shandra, Shandra, and London 2008). We classify forestry statistics as having low reliability if they are based on expert estimates, which often involves extrapolation from an outdated national inventory. As such, we include a dummy variable to measure the reliability of the deforestation data, identifying those nations in which forest cover measures are based upon remote sensing surveys or current national field sampling estimates and are, therefore, of higher quality (1 = high data quality). The reference category includes nations whose forestry estimates are based upon expert estimates or an outdated inventory (0 = low data quality). The coefficients for the data quality measure fail to reach a level of statistical significance.

References

Allen, J. C. and D. F. Barnes. 1985. "The Causes of Deforestation in Developing Countries." *Annals of the Association of American Geographers* 75(2):163–184.

Allison, P. D. 1999. *Multiple Regression: A Primer.* Newbury Park, CA: Pine Forge Press.

Amin, S. 1976. *Unequal Development: An Essay on the Social Transformations of Peripheral Capitalism.* New York: Monthly Review Press.

Austin, K. F. 2010. "The 'Hamburger Connection' as Ecologically Unequal Exchange: A Cross-National Investigation of Beef Exports and Deforestation in Less-Developed Countries." *Rural Sociology* 75(2):270–299.

———. 2012. "Coffee Exports as Ecological, Social, and Physical Unequal Exchange: A Cross-National Investigation of the Java Trade." *International Journal of Comparative Sociology* 53(3):155–180.

Barbosa, Luiz C. 2001. *The Brazilian Amazon Rainforest: Global Ecopolitics, Development, and Democracy*. Lanham, MD: University Press of America.

Bryant, R. L. and S. Bailey. 1997. *Third World Political Ecology*. London: Routledge.

Bunker, Stephen. 1985. *Underdeveloping the Amazon: Extraction, Unequal Exchange, and the Failure of the Modern State*. Chicago, IL: University of Chicago Press.

Bunker, Stephen G. and Paul S. Ciccantell. 2005. *Globalization and the Race for Resources*. Baltimore, MD: Johns Hopkins University Press.

Burns, T. J., E. L. Kick, and B. L. Davis. 2003. "Theorizing and Rethinking Linkages between the Natural Environment and the Modern World-System: Deforestation in the Late 20th Century." *Journal of World-Systems Research* 9(2):357–392.

Butler, R. 2012. *Environmental Impact of Mining in the Rainforest*. Menlo Park, CA: Mongabay.

Clapp, J. and P. Dauvergne. 2005. *Paths to a Green World: The Political Economy of the Environment*. Cambridge, MA: MIT Press.

Dietz, Thomas, R. Scott Frey, and Linda Kalof. 1987. "Estimation with Cross-National Data: Robust and Non-Parametric Methods." *American Sociological Review* 52(3):380–390.

Downey, L., E. Bonds, and K. Clark. 2010. "Natural Resource Extraction, Armed Violence, and Environmental Degradation." *Organization & Environment* 23(4):417–445.

Ehrhardt-Martinez, K., E. M. Crenshaw, and J. C. Jenkins. 2002. "Deforestation and the Environmental Kuznets Curve: A Cross-National Investigation of Intervening Mechanisms." *Social Science Quarterly* 83(1):226–243.

Evans, Peter B. 1979. *Dependent Development: The Alliance of Multinational, State, and Local Capital in Brazil*. Princeton, NJ: Princeton University Press.

Food and Agriculture Organization. 2010. *The United Nations Statistics Division*. Rome: Food and Agriculture Organization.

———. 2015. *The United Nations Statistics Division*. Rome: Food and Agriculture Organization.

Frank, D. J. 1999. "The Social Bases of Environmental Treaty Ratification." *Sociological Inquiry* 69(4):523–550.

Giljum, S. and N. Eisenmenger. 2004. "North–South Trade and the Distribution of Environmental Goods and Burdens: A Biophysical Perspective." *Journal of Environment & Development* 13(1):73–100.

Hornborg, Alf. 2003. "Cornucopia or Zero-Sum Game? The Epistemology of Sustainability." *Journal of World-Systems Research* 9(2):205–218.

Hurst, P. 1990. *Rainforest Politics: Ecological Destruction in South-East Asia*. London: Zed Books.

Jaccard, J. 2001. *Interaction Effects in Logistic Regression*. Thousand Oaks, CA: Sage.

Jorgenson, A. K. 2005. "Unpacking the International Power and the Ecological Footprints of Nations: A Quantitative Cross-National Study." *Sociological Perspectives* 48(3):383–402.

———. 2006. "Unequal Ecological Exchange and Environmental Degradation: A Theoretical Proposition and Cross-National Study of Deforestation, 1990–2000." *Rural Sociology* 71(4):685–712.

———. 2010. "World-Economic Integration, Supply Depots, and Environmental Degradation: A Study of Ecologically Unequal Exchange, Foreign Investment Dependence, and Deforestation in Less-Developed Countries." *Critical Sociology* 36(3):453–477.

———. 2012. "The Sociology of Ecologically Unequal Exchange and Carbon Dioxide Emissions, 1960–2005." *Social Science Research* 41(2):242–252.

Jorgenson, A. K. and T. J. Burns. 2007. "Effects of Rural and Urban Population Dynamics and National Development on Deforestation in Less-Developed Countries, 1990–2000." *Sociological Inquiry* 77(3):460–482.

Jorgenson, A. K., K. Austin, and C. Dick. 2009. "Ecologically Unequal Exchange and the Resource Consumption/Environmental Degradation Paradox: A Panel Study of Less-Developed Countries, 1970–2000." *International Journal of Comparative Sociology* 50(3–4):263–284.

Li, Q. and R. Reuveny. 2006. "Democracy and Environmental Degradation." *International Studies Quarterly* 50(4):935–956.

London, Bruce. 1987. "The Structural Determinants of Third World Urban Change: An Ecological and Political Economic Analysis." *American Sociological Review* 52(1):454–463.

London, B. and R. J. S. Ross. 1995. "The Political Sociology of Foreign Direct Investment: Global Capitalism and Capital Mobility, 1965–1980." *International Journal of Comparative Sociology* 36(3):198–219.

London, B. and D. A. Smith. 1988. "Urban Bias, Dependency, and Economic Stagnation in Non-Core Nations." *American Sociological Review* 53(3):454–463.

Marshall, M. G. and K. Jaggers. 2006. *Polity IV Country Reports*. Available at www.systempeace.org.

Midlarsky, M. I. 1998. "Democracy and the Environment: An Empirical Assessment." *Journal of Peace Research* 35:341–361.

Moore, S., A. C. Teixeira, and A. Shiell. 2006. "The Health of Nations in a Global Context: Trade, Global Stratification, and Infant Mortality Rates." *Social Science and Medicine* 63(1):165–178.

Muradian, R. and J. Martinez-Alier. 2001. "Trade and the Environment from a 'Southern Perspective'." *Ecological Economics* 36(2):281–297.

O'Connor, J. R., editor. 1998. *Natural Causes: Essays in Ecological Marxism.* New York: Guilford Press.

Rich, Bruce A. 1994. *Mortgaging the Earth: The World Bank, Environmental Impoverishment, and the Crisis of Development.* Boston, MA: Beacon Press.

Roberts, J. T. and B. C. Parks. 2007. "Fueling Injustice: Globalization, Ecologically Unequal Exchange, and Climate Change." *Globalizations* 4(2):193–210.

Rudel, Thomas K. 1989. "Population, Development, and Tropical Deforestation: A Cross-National Study." *Rural Sociology* 54(3):327–338.

———. 2013. "The National Determinants of Deforestation in Sub-Saharan Africa." *Philosophical Transactions of the Royal Society B* 368(1625):20120405.

Rudel, Thomas K. and J. Roper. 1997. "The Paths to Rain Forest Destruction: Cross-National Patterns of Tropical Deforestation, 1975–1990." *World Development* 25(1):53–65.

Schofer, E. and A. Hironaka. 2005. "The Effects of World Society on Environmental Protection Outcomes." *Social Forces* 84(1):25–47.

Shandra, J. M. 2007. "Economic Dependency, Repression, and Deforestation: A Quantitative, Cross-National Analysis." *Sociological Inquiry* 77(4):543–571.

Shandra, J. M., C. Leckband, L. A. McKinney, and B. London. 2009a. "Ecologically Unequal Exchange, World Polity, and Biodiversity Loss a Cross-National Analysis of Threatened Mammals." *International Journal of Comparative Sociology* 50(3–4):285–310.

Shandra, J. M., C. Leckband, and B. London. 2009b. "Ecologically Unequal Exchange and Deforestation: A Cross-National Analysis of Forestry Export Flows." *Organization & Environment* 22(3):293–310.

Shandra, J. M., E. Shor, and B. London. 2009c. "World Polity, Unequal Ecological Exchange, and Organic Water Pollution: A Cross-National Analysis of Developing Nations." *Human Ecology Review* 16(1):53–63.

Shandra, J. M., C. L. Shandra, and B. London. 2008. "Women, Non-Governmental Organizations, and Deforestation: A Cross-National Study." *Population and Environment* 30(1–2):48–72.

Shandra, J. M., E. Shircliff, and B. London. 2011. "The International Monetary Fund, World Bank, and Structural Adjustment: A Cross-National Analysis of Forest Loss." *Social Science Research* 40(1):210–225.

Smith, Jackie and Dawn Wiest. 2005. "The Uneven Geography of Global Civil Society: National and Global Influences on Transnational Association." *Social Forces* 84(2):621–652.

Snow, K. H. 2015. "Foreign Mining, State Corruption and Human Rights in Mongolia." *Global Research* (May 22). Available at www.globalreach.ca.

Tabachnick, Barbara G. and Linda S. Fidell. 2013. *Using Multivariate Statistics* (Sixth Edition). Boston, MA: Pearson.

United Nations Commodity Trade Statistics Database. 2008. UN Comtrade. Available at www.comtrade.un.org.

———. 2015. UN Comtrade. Available at www.comtrade.un.org.

———. 2016. UN Comtrade. Available at www.comtrade.un.org.

World Bank. 2003. *World Development Indicators*. Washington, DC: World Bank.

———. 2005. *World Development Indicators*. Washington, DC: World Bank.

———. 2015. *World Development Indicators*. Washington, DC: World Bank.

World Resources Institute. 2005. *World Resources Institute Global Forest Watch. Forests*. Washington, DC: World Resources Institute.

World Wildlife Fund. 2016. *Deforestation Threats*. Washington, DC: World Wildlife Fund.

York, Richard, Eugene A. Rosa, and Thomas Dietz. 2003. "Footprints on the Earth: The Environmental Consequences of Modernity." *American Sociological Review* 68(2):279–300.

8

From Sea Slaves to Slime Lines: Commodification and Unequal Ecological Exchange in Global Marine Fisheries

Brett Clark, Stefano B. Longo, Rebecca Clausen, and Daniel Auerbach

Few global food industries are as free from the public view as marine fisheries. According to the Millennium Ecosystem Assessment (2005:61), "fishing is the most direct anthropogenic force affecting the structure, function, and biodiversity of the oceans." In fact, every square mile of the World Ocean is affected by human activities, such as overfishing, nutrient runoff, acidification, and pollution (Halpern et al. 2008). Fortunately,

B. Clark (✉) • D. Auerbach
Department of Sociology, University of Utah, Salt Lake City, UT, USA
e-mail: brett.clark@soc.utah.edu; daniel.auerbach@soc.utah.edu

S. B. Longo
Department of Sociology and Anthropology, North Carolina State University, Raleigh, NC, USA
e-mail: Stefano_longo@ncsu.edu

R. Clausen
Department of Sociology and Human Services, Fort Lewis College, Durango, CO, USA
e-mail: clausen_r@fortlewis.edu

R. S. Frey et al. (eds.), *Ecologically Unequal Exchange*,
https://doi.org/10.1007/978-3-319-89740-0_8

there are scholars and activists who are examining the decline in fish stocks due to overfishing and inexcusable levels of bycatch (Pauly 2009; Roberts 2012). Nevertheless, the dynamics of the capitalist food system, including human rights violations against those who work in the global fisheries, often remain hidden from the public.

In a rare exposé, the *New York Times* published an important five-part series, titled "The Outlaw Ocean," documenting widespread violence and crimes on the high seas. The articles exposed extensive environmental degradation, violations of fisheries regulations, and labor injustices that are connected to producing food. Two of the five articles focused specifically on the conditions of "slave labor" aboard fishing boats in Southeast Asia (Urbina 2015a, 2015b). Additionally, the Asia Foundation and International Labor Organization (2015) revealed that the labor abuses associated with the fishing industry do not end once the boats reach shore. For example, in land-based seafood processing plants, migrant children are subjected to an array of life-threatening occupational hazards. It was determined that extreme forms of labor exploitation are entrenched within global marine fisheries, which connect the global South and North.

The popular media has provided a great service with its investigative journalism on this topic. Yet, the explanations for such conditions are cursory at best. It has been noted that environmental and labor abuses occur primarily because the oceans are so vast, jurisdiction is complicated due to the lack of private property rights, and policing is rare. A particular explanatory emphasis is put on the corruption of politicians and law enforcement agencies, as well as rogue criminal actors. While all of these points are true, they do not address the factors that have influenced the organization of global fisheries and created the conditions for the proliferation of associated problems. In contrast, we argue that the capitalist system, as the social organization of generalized commodity production, has established distinct patterns of expropriation of marine species and fishery workers, which is in part built upon the unequal economic and ecological exchange between the global South and North. These dynamics have influenced fishing operations at sea and the processing of fish on land, contributed to overfishing, led to the development of environmentally damaging aquaculture operations in Southeast Asia, and fostered global social inequality and environmental injustice. For our analysis, we draw on both the tragedy of the commodity and unequal ecological exchange (sometimes called "ecologically unequal exchange") theories. Thus, we begin with a discussion of each of these perspectives. We then present brief case studies, moving from "sea slaves" to "slime lines," highlighting how the global division of labor associated with marine commodity production creates specific patterns of exploitation, degradation, and unequal exchange.

The Tragedy of the Commodity

The historic decline in global fish stocks, especially during the twentieth century, generated much interest among economists and natural resource managers to explain why this was occurring and what could be done to address this concern (Gordon 1954; Schaefer 1957; Scott 1955). In the 1960s, drawing upon various insights from economists, ecologist Garret Hardin proposed an analysis asserting that a combination of overpopulation and greedy individuals will eventually destroy resources that are held in common, such as fisheries. For him, individuals are inherently driven by self-interest, which results in the exploitation of the commons without regard for the potential social or ecological consequences. This conception serves as the basis of his theory of the "tragedy of the commons" (Hardin 1968). Following this approach, there are two broad options to prevent this tragedy from occurring: (1) a top-down system of state control of common property resources could be implemented, which would regulate the commons; or, (2) the commons could be converted into a system of private ownership, which would limit access and encourage protection of resources. Hardin principally favored the second option, but scientists and policy makers following this logic have generally proposed both. While a variety of approaches have been employed to try to manage fisheries, the most pervasive explanation for fisheries depletion or collapse in scientific and policy circles remains the tragedy of the commons thesis.

We find the tragedy of the commons perspective to be sociologically imprecise, and thus problematic. It tends to lack historical insights and social context, stressing individual behavior, and it focuses on the wrong aspect of social organization, the commons, as the source of the problem (Cox 1985). As a counter, we propose that the "tragedy of the commodity" perspective provides a more appropriate analytical framework to assess the history of fisheries under capitalist development (see Longo, Clausen, and Clark 2015). Here we briefly discuss the logic of capital and commodification, in order to outline this approach and to set up some of the dynamics that contribute to unequal economic and ecological exchange within the global food system.

Capitalism is a system of generalized commodity production, premised on the endless accumulation of capital. As a grow-or-die system, it must expand exchange value, which is purely a quantitative measure (Burkett 1999). Economist Robert Heilbroner (1985:36) explained that capital is best understood as the "continuous transformation of capital-as-money into capital-as-commodities, followed by a retransformation of capital-as-commodities into capital-as-more-money." This growth dynamic is viewed by capital as limitless.

Karl Marx (1976) examined the one-sidedness of the value form in capitalism. Capital treats nature as a "free gift." At the center of Marx's critique of capitalism is the contradiction between wealth (which he argued includes nature as well as labor) and value (which capital views as based on labor alone). Within Marx's "ecological value-form analysis," nature constituted, along with labor, one of the two sources of wealth. For capital, nature was not part of the value calculus, facilitating the destructive tendencies associated with endless accumulation (Burkett 1999; Marx 1976:745).

From the tragedy of the commodity perspective, the commodity serves as the basic unit for understanding the larger culture–nature relations of capitalism. An assessment of the commodity and its social relations serves as a foundation for analyzing production, exchange, distribution, and consumption within the capitalist regime. All of these points are part of a definite process. But in a society organized around the dictates of capitalist markets, production is the point of departure, as it is broadly geared to facilitate capital accumulation. Within this socioeconomic system, a commodity is a product sold on a market, produced for exchange rather than for use. As Marx (1976) described, during the production process, humans interact with the larger biophysical world, directly or through the labor accumulated within technology (e.g., machines). The growth imperative of capital subordinates labor and nature to the impulse toward further accumulation. It shapes the division of labor, labor conditions, and the material exchanges with ecological systems.

Under this system, money dominates the organization of socioecological relationships. Marxian economist Paul Burkett (1999:64–65) notes that the unity between social production and nature is mystified in the operation of capital, by the "increasing domination of exchange value over use value." Money serves as the vehicle by which products are taken out of any ecosystem and productive relationship, simply to be exchanged within the global market. The representative value of money is abstracted from the intricate diversity and subtleties of nature. This results in capitalism imposing its logic, often times through the simplification and transformation of natural systems to further accumulation. Additionally, the ceaseless drive to accumulate generally intensifies production operations, increasing demands placed on the biophysical world—whether in the form of fish harvested from the seas, as resources used in the production of commodities, or as waste generated from this process (Longo et al. 2015; Schnaiberg 1980; Sweezy 2004). Such actions usually disregard natural cycles, systems, and life cycles that are not immediately relevant to the logic of accumulation, and thus have often resulted in an array of ecological contradictions and forms of environmental degradation (Dickens 2004; Foster 1994, 2000; Foster, Clark, and York 2010; Longo et al. 2015, 2016; Sanderson and Frey 2014).

The ability of capital to extract surplus value through commodity production is made possible through the exploitation of human labor and the expropriation of nature. In other words, capital does not pay the true costs of production. The veil of commodity production hides the fact that both labor and nature provide the basis of wealth, preventing a systemic assessment of the social relations driving environmental degradation. The tragedy of the commodity approach helps lift the veil to reveal the social–natural relations of capitalist commodity production that contribute to expropriation, the accumulation of capital, and environmental degradation. By shifting the emphasis from individuals toward the systemic dynamics of commodification, the approach also illuminates the social forces that contribute to unequal exchanges throughout the global capitalist system.

Unequal Economic and Ecological Exchange

The theory of unequal ecological exchange has been used to examine the material–ecological flows that transform socioeconomic and environmental relations within and between nations (Bunker 1984; Frey 1994; Hornborg 2003). The movement of economic values shadows in complex ways the vertical transference of matter and energy throughout the international hierarchy of nations, while producing distinct consequences regarding social inequalities and environmental degradation (Foster and Holleman 2014; Hornborg 1998; Jorgenson 2006; Rice 2009). Control of such economic and material flows is central to the forces of competition and the accumulation of capital. As a result, unequal economic exchange and unequal ecological exchange are intertwined.

Analysis of unequal exchange accompanied the rise and expansion of capitalism. Classical political economists, such as Adam Smith (1937), John Stuart Mill (1877), and David Ricardo (1951), highlighted that colonialism and the plunder of resources from what is now known as the global South played an important role in constituting capitalism as a system. For Marx (1976:926), "capital comes [into the world] dripping from head to toe, from every pore, with blood and dirt." In large part, this is due to the primary accumulation process, which helped establish divisions between the core and periphery of the capitalist system, as the wealth of distant lands was expropriated through various mechanisms, including slavery, the enclosure of commons, destruction of communities and home markets, and the global division of labor. Marx (1976:915–918) drew out these connections, observing "the discovery of gold and silver in America, the extirpation, enslavement and entombment in mines of the indigenous population of that continent, the beginnings of the conquest and plunder of India, and the conversion of Africa into a preserve for the commercial hunting of blackskins, are all things which characterize the dawn of the era of capitalist production. These idyllic proceedings are the chief moments of primitive accumulation.... The treasures captured outside Europe by undisguised looting, enslavement and murder flowed back to the mother-country and were turned into capital there."

Marx analyzed how the creation and operations of the global capitalist system facilitated the unequal exchange between nations, whereby distant labor, lands, and ecosystems became mere appendages to the growth requirements of the "advanced" center. The extensive division of labor and extreme forms of labor exploitation created a situation where "the profit rate is generally higher there [i.e., colonies] on account of the lower degree of development, and so too is the exploitation of labour, through the use of slaves and coolies, etc." (Marx 1991:345). These historical conditions set up a situation where through international trade, "the privileged country receives more labour in exchange for less, even though this difference, the excess, is pocketed by a particular class, just as in the exchange between labour and capital in general. Thus in as much as the profit rate is higher because it is generally higher in the colonial country, favourable natural conditions there may enable it to go hand in hand with lower commodity prices" (Marx 1991:345–346). Marx clarified that wealth, which includes both nature and labor, is transferred from the global South to the North, and accumulated by the capitalist class, as part of the unequal exchange process associated within the structure of the capital system. Contrary to the notion of comparative advantage, where trade can result in both nations benefitting, Marx notes that this process often deepens global inequalities between nations, since "two nations may exchange according to the law of profit in such a way that both gain, but one is always defrauded.... Not only individual capitalists, but also nations may continually exchange with one another, may even continually repeat the exchange on an ever-expanding scale, without for that reason necessarily gaining in equal degrees. One of the nations may continually appropriate for itself a part of the surplus labour of the other, giving back nothing for it in the exchange" (Marx 1993:872).

In the 1960s and 1970s, as both dependency and monopoly capital theories were being further developed, unequal exchange relations were running concerns. Paul Baran and Paul Sweezy (1966) argued that the ongoing transfer of wealth from the periphery to the core stemmed from the unequal relationship between nations, which created terms of trade that favored the global North. Development projects that encouraged specialization in raw materials extraction and/or cash crops furthered the net transfer of surplus to the core (Frank 1967; Grivan 1976; Guevara

1997; Magdoff 1978). Eduardo Galeano (1997:2) explained that "Latin America is the region of open veins. Everything, from the discovery until our times, has always been transmuted into European—or later United States—capital, and as such has accumulated in distant centers of power. Everything: the soil, its fruits and its mineral-rich depths, the people and their capacity to work and to consume, natural resources and human resources." He goes on to note "the more a product is desired by the world market, the greater the misery it brings to the Latin American peoples whose sacrifice creates it" (Galeano 1997:61). Arghiri Emmanuel (1972) proposed that unequal exchange between nations was rooted in the international mobility of capital and the immobility of labor, which resulted in the transfer of "hidden" value from low-wage to high-wage countries. Samir Amin (1976, 2010) and John Smith (2016) indicate that as modern technology is being deployed in production throughout the world, global wage inequalities have served as an important foundation for the transfer of value created in the global South to the North.

Unequal ecological exchange suggests that there is an "exchange of more ecological use value (or nature's product) for less" (Foster and Holleman 2014:205). In general, it is argued that there is a "vertical flow of exports" from the global South to the North, whereby the latter partially externalize their consumption-based environmental costs, which increases the concentration of environmental degradation in the former (e.g., Bunker 1984; Chase-Dunn 1998; Frey 1994; Hornborg 1998, 2011; Jorgenson 2006; Jorgenson and Clark 2009; O'Connor 1998; Roberts and Parks 2007; Srinivasana et al. 2008). The populations of the global North are historically positioned more advantageously in the world-economy, and thus are able to secure and maintain favorable terms of trade allowing for greater access to the natural resources and sink capacity of bio-productive areas within the South. As a result, the countries in the North over-utilize global "environmental space," which encompasses the stocks of natural resources and waste assimilation properties of ecological systems supporting human social organization (Rice 2007, 2008). The misappropriation of environmental space suppresses resource consumption opportunities for many countries in the global South, which also negatively affects the well-being of their domestic populations (Jorgenson, Austin, and Dick 2009).

Unequal ecological exchange analysis has helped reveal the environmental and social inequalities that have accompanied capitalist development. Stephen Bunker and Paul Ciccantell (2005) argue that the central feature of the capitalist interstate system has been the systemic exploitation of nature through the international division of labor. As a result, the North has been able to over-utilize "environmental space" (Rice 2007). Many consumption-based environmental impacts of nations have been externalized to the South, such as through deforestation, biodiversity loss, water pollution, dumping of toxic waste, and carbon dioxide emissions (e.g., Frey 1994, 2015; Jorgenson 2006; Jorgenson and Clark 2011; Lawrence 2009; Rice 2009; Roberts and Parks 2007; Shandra et al. 2009a; Shandra, Shor, and London 2009b; Stretesky and Lynch 2009). Recent work demonstrates the material–ecological transfer from the global South to the North (Foster and Holleman 2014). While this approach is still developing in innovative ways, it is clear that it helps highlight additional dimensions of the social and ecological transfers and inequalities structured by the global capitalist system.

Sea Slaves

In the contemporary period, slavery is also referred to as "forced labor." While the terms are not always legally identical, they are often used interchangeably. The act of owning human beings as property is officially illegal everywhere in the world. Regardless of whether formally considered slaves, forced laborers, or debt-peons, these individuals are coerced to work, often physically restrained, and endure cruel conditions, including mental and physical abuse or threat of abuse (International Labor Organization 2015). The process through which individuals are brought into these conditions is often called "human trafficking."

It is estimated that approximately 21–27 million people around the world are victims of human trafficking (International Labor Organization 2013; U.S. Department of State 2015). While it is well known that the majority of individuals bonded into slavery are forced into sex-work, there is growing awareness that a large number of people who are trafficked into forced labor end up in the global fishing sector (Bales 2016;

UNIAP 2009). There is a long history of transporting slaves using ships. Today, "sea slaves" are still bound to ships, harvesting, via their labor, valuable commodities that are transferred to distant markets.

The actual numbers of individuals who are slaves in the commercial fishing sector is unknown. According to some reports, 145,000–200,000 individuals are enslaved just in Thailand-based fishing operations (Irvine, Mohsin, and Olarn 2015). Fishing remains a domain where slavery is relatively easy to impose due to its existence in the marine realm. In many ways, it is largely out of sight or socially invisible. The global fishing sector has been notoriously difficult to regulate. Even under established legal arrangements, labor practices at sea are some of the most dangerous working conditions and illegal fishing is common.

There has been increasing discussion of the labor conditions in the fishing sector in the South China Sea (International Labor Organization 2013). Recent journalist exposés point specifically to the widespread use of slave labor in Thailand's fishing sector, but Thailand is not the only nation where such labor is utilized in seafood operations. People have been trafficked or exploited in forced labor conditions in the fishing sectors of other Asian and West African nations (International Labor Organization 2013, 2015; Urbina 2015a).

In the circulation of capital, labor is key in so far as adding value in the production of commodities. To maximize profits, capital attempts to minimize wages and other expenses. As Marx (1991) noted, the potential profits from the global South were often higher due to extreme forms of labor exploitation. Global wage inequalities, especially in regard to slave labor, underlie the unequal economic and ecological exchange (Amin 2010; Foster and Holleman 2014). Slave labor at sea and poor working conditions support an extensive global food system, whereby profits accumulate along the commodity chain, with little to no wages spent on workers at the point of harvest. The goal on these boats is to maximize harvests, which affects the population of fish, through what is known as the super-exploitation of labor. Many of the operations that use forced labor are likely engaged in other unlawful activities, such as illegal fishing, which is a rampant problem for the global fishing sector, as it faces serious concerns related to fisheries depletion from overfishing and other ecosystem impacts.

The conditions for these laborers are, of course, dreadful, which results in minimal expenses on the part of capital. Those individuals who are lucky enough to escape from the ships have described the conditions and circumstances. Sea slaves are forced to work long hours with little sleep. In order to evade the authorities, fishing operations that utilize slave labor sometimes stay at sea for years without returning to port. The fishing boats regularly meet with "mother ships" to offload captures, relieve captains, and receive supplies. When the fishing boats do return to shore, laborers are often forcefully secured to the boat via chains and other harnesses (International Labor Organization 2013). These conditions at sea make labor more easily exploitable as interactions outside the immediate individuals on the boat are kept to a minimum.

Physical abuse is common. Individuals are beaten for working too slowly and/or making mistakes, such as placing the wrong type of fish in a bucket meant for a different species (Urbina 2015b). Workers are often sold to other fishing operations, only to experience similar conditions. Slaves who are caught trying to escape are subject to beatings and even murdered—since they can be replaced under current conditions. Some of these workers die from illness, overwork, or drowning during attempts to escape. In order to maximize productivity, workers are frequently given amphetamines to increase their energy, allowing them to work more hours during the day (up to 20 hours). While the labor conditions create illness and stimulant drugs are provided, curative medicines are not readily available. Instead the workers suffer with sickness and injuries—such as the loss of fingers severed by nylon ropes and nets. The fishing boats are relatively small, so when workers are allowed to sleep, they rest in very cramped quarters (Environmental Justice Fund 2014; UNIAP 2009).

These working conditions are linked to larger structural changes in the international division of labor and the global economy of commodity production. Lacking access to the means of production, general trends of overproduction, unemployment, and falling wages, individuals become desperate for work. Many of the people who end up on fishing ships as slaves initially sought out employment agencies to place them in jobs abroad. For example, a large number of Filipino workers have emigrated for work in the fishing sector throughout Asia, only to end up being trafficked into slave labor (Urbina 2015a). In the case of Thailand's fishing sector, poor Thai laborers and migrants from nearby nations such as Cambodia, Myanmar, Laos, and Indonesia have been the principal sources of slave labor (U.S. Department of State 2015). Oftentimes, individuals are promised work abroad in factories or in agriculture, but end up being sold to fishing operations. When they arrive, their identification documents are confiscated. They are immediately indebted to the brokers and boat captains for broker's fees and recruitment costs, and are kept in ongoing debt bondage for many years (U.S. Department of State 2015; UNIAP 2009).

Fisheries using slave labor are not marginal operations within the global economy. Thailand is the third largest exporter of seafood commodities, behind China and Norway, with exports valued at about $7 billion annually (United Nations 2014). The United States and the European Union are the largest importers of fish from Thailand. The fish captured by slave labor is not always directly exported for human consumption, as it also provides an array of resources used throughout the commodity chain and in products destined for the global North. The connections of slave labor to other commodities and environmental degradation are evident when considering marine production in general.

Thailand's capture fishing sector has experienced the challenge now commonly associated with depleted fisheries. Large-scale industrial fishing operations have been introduced to maximize harvests. The sole focus on maximizing profit, and the general disregard for environmental conditions and life cycles, has led to a situation where fish are being captured at a rate faster than they can reproduce. As more fish are extracted at an increasingly faster pace from marine systems, the populations of desired species are plummeting. The Food and Agriculture Organization of the United Nations estimates that approximately two-thirds of the world's fisheries are either overfished or fully fished (United Nations 2014). The bounty of the oceans is being diminished through commodity production systems, as profits accumulate along the commodity chain.

According to a report by the Environmental Justice Fund (2014), using data from the Department of Fisheries in Thailand, the fish catch on Thai fishing boats have been in decline over the last several decades. An increasing effort is required to capture fish in the surrounding waters including the Gulf of Thailand, Andaman Sea, and the South China Sea. In fisheries science, this is measured as catch per unit effort (CPUE), which is a calculation of the type and amount of capture practices in relation to the total numbers or weights of captured fish. It provides an indication of the relative abundance of fish populations and the energy and capacity required to capture fish. A decreasing CPUE can be an indication of a reduction in available fish, as they become scarcer and harder to capture. In the case of Thailand, it has been estimated the there was a 97 percent reduction in CPUE in the Gulf of Thailand since the 1960s (Environmental Justice Fund 2014). Thus, fishing boats in this part of the world are paradoxically increasing their capacity (size), energy, and time fishing, which only exacerbates the decline in captures. Such conditions create economic problems, as the expenses of operating a fishing boat increase due to the time at sea, the distance that must be traveled to capture fish, and the types of boats required for such activities. Due to the dynamics of global fisheries markets, including the growth of aquaculture, the fastest growing food sector in the world, fishing boats are not necessarily able to reap increased returns on scarcer captures. Under these conditions, exploiting slave labor has served as a way to minimize labor costs in an effort to maintain profitability.

A result of these changing marine and economic circumstances is that large portions of the captures in Thailand's slave-labor fishing boats are so-called trash fish—a catch-all term referring to sea life that have little or no direct market value, unless transformed into a different commodity. Non-targeted fish, such as bycatch or trash fish, are usually processed and turned into fishmeal and sometimes fish oil, which is used in products like animal feed and fertilizer. The capture of non-targeted fish has long had ecological consequences, exhausting a broad range of marine species, on which many people throughout the world depend (Clausen and Clark 2005).

The leading seafood commodity exported from Thailand is shrimp. In 2009, Thailand exported over $2.7 billion in shrimp products (United Nations 2011). Although there is some small-scale capture of shrimp in Thailand, a large portion of export shrimp from Thailand is reared in aquaculture facilities. Of the almost 600,000 tons of shrimp and prawns produced in Thailand in 2010, 90 percent was farmed (United Nations 2011). Thus, the vast majority of the exported shrimp is not directly captured by slave labor. Nevertheless, the connection between slavery at sea and farmed production of seafood remains strong when considering the historical developments within the larger commodity sector.

Given the declining fish population of targeted fish, trash fish are increasingly seen as means to supply needed resources for other commodity lines. The largest consumers of fishmeal are aquaculture and livestock operations. In particular, shrimp aquaculture facilities use fishmeal and fish oil as inputs. Following the commodity chain of global food production, there is a direct link between the fishing operations making use of slave labor and the shrimp produced in aquaculture facilities for wealthy markets in the United States and Europe. The exploitation of slave labor and expropriation of fish populations underlies seafood commodity production, and it is not limited to commodities destined for human consumption.

Another major buyer of processed trash fish is the pet food industry. Recent investigative reporting following this commodity chain has determined that leading producers of pet food in the United States and Europe, such as Nestlé, use fish captured by ships using slave labor. Thus, pets in the wealthiest part of the world are fed products of slave labor that are increasing the pressures placed on fish populations. According to a recent *New York Times* report, "The United States is the biggest customer of Thai fish, and pet food is among the fastest growing exports from Thailand, more than doubling since 2009" (Urbina 2015b). Sales of fish for pet food totaled over $190 million in 2014. "The average pet cat in the United States eats 30 pounds of fish per year, about double that of a typical American" (Urbina 2015b). The pets of the global North also thrive on the slavery, humiliation, deprivation, and murder of the poorest workers in the world. All the while, capital accumulates along the commodity chain, transferring wealth through unequal economic and ecological exchange, expropriating environmental space within the global South, as global inequality deepens and the ecological conditions in the oceans worsen.

Slime Lines

The capture of fish is often glamorized, such as in literary works like *The Old Man and the Sea* and *Moby Dick* (Muszynski 1996). Contemporary versions of this narrative can be found in reality shows on television, such as *The Deadliest Catch*. These romantic, often masculine, notions of fishing in the open ocean are far from the reality of the international division of labor and the forced fishing labor described above. Just as concerning, however, is how pervasive commodity fetishism is in regard to this industry. The laborers who do the remainder of the work to prepare fish for sale after capture are largely invisible. Thus, it is important to note that labor abuses do not end when the fishing boats reach the shore or transfer their catch. Once fish have been captured from the ocean, or harvested from aquaculture facilities, they are often delivered to a land-based processing plant where a different population of workers are required to prepare the seafood for global markets.

Seafood processing plants are sometimes referred to as "slime lines" due to the factory-like design of dead fish moving down conveyor belts, passing through the hands of multiple workers involved in single-skill tasks. In this highly Taylorized operation, each worker is responsible for one part of the disassembly of the marine species, such as heading, gutting, skinning, filleting, peeling, cleaning, and sorting (Jeebhay, Robins, and Lopata 2004). Not all processing plants use the same design, and the worker's task will of course vary by the species being butchered, but the extensive division of labor associated with mass production, standardization, and a deskilled labor force is fairly common throughout the seafood processing sector. With this industrial design arises similar potentials for labor exploitation, as is seen in other sectors, such as garment manufacturing. Here we briefly highlight the worker abuses associated with seafood processing plants, including child labor, occupational hazards, and reliance on vulnerable migrant laborers. In the effort to maximize profits, capital seeks a cheap labor force, resulting in gross forms of social inequality and both unequal economic and ecological exchange.

The International Labor Organization defines child labor as work in conditions that "deprive children of their childhood, their potential, and their dignity, and that [are] harmful to physical and mental development" (Asia Foundation and International Labor Organization 2015:41). It refers to work that is mentally, physically, socially, or morally dangerous and harmful to children, requiring them to leave school early or attempt to combine school attendance with excessively long and heavy work. The International Labor Organization (2015) estimates that 168 million children between the ages of 5 and 17 currently work under conditions that are considered illegal, hazardous, or extremely exploitative. Underage children work at various jobs around the world, usually because they and their families are extremely poor. By far, agriculture and fishing industries combined have the highest incidence of child labor, with 59 percent of children 5–17 years old, who are working, employed in these sectors.

A recent investigation presents qualitative data documenting the extent of child labor in the marine seafood processing plants of Thailand. There are currently 185 registered shrimp processing facilities in Thailand. Seventeen of these are large-scale processing sites, while the majority are small- and medium-sized facilities. The Asia Foundation and the International Labor Organization collaborated to conduct focus group discussions and interviews with child workers and their families in various Thai communities that host seafood processing plants. The goal of the project was to gain better and more detailed information regarding child labor and labor conditions within seafood processing. The findings are that a high proportion of children working in the seafood industry fell under the International Labor Organization definition of child labor—36 percent of children in one province and 40 percent of children in another. On average, one in three children employed in seafood processing are not attending school, with one province reporting as high as 79 percent of the child workers with no regular school attendance (Asia Foundation and International Labor Organization 2015:127). Among non-school-going children in the seafood processing industries, the main reason given for non-attendance is financial, with 56 percent of children needing to work for family income. Given the lucrative nature of the shrimp industry with profitable exports to restaurants around the world, it is evident that this commodity sector, with its low wages, results in the exchange of more labor and ecological use value for less within the global economic system.

In addition to forgoing an education, children working in the seafood processing industries are more frequently exposed to occupational hazards than those who work in other industries, which create distinct patterns of environmental dangers and risks. In an effort to minimize costs, there is a tendency to avoid investment in worker safety precautions or appropriate handling of waste and hazardous outputs, which helps deepen the consequences of unequal exchange. Approximately 26 percent of children work directly with fire, gas, or flames, and 23 percent work in wet and dirty conditions (Asia Foundation and International Labor Organization 2015:16). These hazardous conditions result in children who work in seafood processing plants being twice as likely to be injured in comparison with other industries. These children also work very long hours in comparison to other jobs. Many of the children report working between the hours of 10:00 p.m. and 6:00 a.m. In comparison to other industries, shrimp and seafood plants have three times the proportion of children working during the night (Asia Foundation and International Labor Organization 2015:110).

The risks of working in these seafood processing plants are not limited to children. Processing plants pose safety risks such as mechanical and electrical accidents, excessive noise, low temperatures, bacterial and parasitic infections, and bioaerosol exposure containing allergens and toxins. The health impacts resulting from these workplace hazards are numerous, including cuts, lacerations, skin infections, sepsis, asthma, and cumulative trauma disorders (Jeebhay et al. 2004). The inhalation of wet aerosols is a less obvious but quite dangerous form of occupational exposure, occurring when workers breathe in the spray from the heading and gutting stations. Additionally, inhalation of formalin and hydrogen sulfide gas from decomposing fish is a health concern. The intensification and expansion of production in this commodity sector has led to more frequent reporting of occupational health problems, suggesting that the need for safety equipment and redesign of workplaces has not been addressed. The global organization of seafood production results in disproportionate health and environmental burdens being placed on vulnerable populations, extending from sea slaves to children to workers in general.

Industrialized fishing has a long history of exploiting workers in order to expand profits. This pattern is just as evident in the global North as it is in Thailand, demonstrating the exploitive nature of the marine fishing industry as part of the larger capitalist food system. Worker abuses extend beyond the use of child labor and hazardous workplaces, and include the exploitation of vulnerable migrant labor populations who provide cheap labor for the industry. Workers, in addition to the seafood, are considered nothing more than a commodity to be bought at the cheapest price and replaced when no longer deemed useful. For example, the early salmon processing plants of the Pacific Northwest used Chinese immigrant labor to reduce labor costs. Patricia Roy (1989) documents how owners of the processing plants preferred Chinese labor because this population, due to discrimination in other parts of the labor market, had to accept the low pay, unpleasant working conditions, and uncertainties of the short season. Chinese labor was heavily relied upon in the salmon processing plants until mechanization and the invention of the "Iron Chink" that replaced some of the human labor requirements. The Iron Chink was a racist name for a machine that displaced Chinese human labor, and gutted and cleaned salmon at a rate comparable to the work of 30 to 40 skilled workers (Radke and Radke 2002). These transformations resulted in the exploitation of labor and the increasing intensity of production, which accompanied expanding the market for seafood at the expense of fish populations and the well-being of workers.

Domestic examples in the United States highlight common patterns throughout the capitalist system, where commodity production allows for distinct conditions of accumulation and exploitation. The invisibility and struggle of migrant populations used for salmon processing continue in that industry today. In Oregon, Washington, and Alaska, Chinese labor began to decline in the 1950s. They were replaced with Mexican and Filipino workers (U.S. Bureau of Labor Statistics 2012). As in other industries, undocumented workers with limited language skills are a vulnerable population, forced to work long hours and afraid to complain for fear of being fired.

The seafood processing plants in Alaska implemented a new way to depress wages through the use of temporary work visas. The Summer Work Travel program, designed by the U.S. Department of State, provides J-1 visas to more than 100,000 college students from around the world, allowing them to work for three months in the United States and then to travel for one month. These workers are exempt from Social Security, Medicaid, and federal unemployment taxes, which saves companies about 8 percent of the wages per person. Processing plant owners, who depend on seasonal work in the summer, are eager to reduce payroll costs by hiring these international workers. In addition, employers do not have to pay health insurance since the participants must pay for it themselves. This plan depresses wages because it gives employers access to workers from other countries who are eager to accept any wage and who are already in debt for the several thousand dollars in fees, travel costs, and so on, for the program. Many of these workers are virtually indentured to the processing owners and are unable to challenge low pay and poor working and housing conditions in rural Alaskan fishing towns.

The J-1 visa exchange structure was established in 1961, and has grown dramatically in the past decade. From 2000 to 2001, the number of J-1 visas nearly doubled from 56,000 to 103,000 with top sending countries being Russia, Ukraine, Turkey, Brazil, and Thailand (Kammer 2011). The recruitment process is blunt, with one website for the Kazakhstan Council for Educational Travel (http://kcet.kz/ru/gray) explaining that the seafood processing job requires a willingness to "work 16 hours a day, seven days a week for cleaning, cutting, and packaging of frozen fish… able to endure four months of harsh climate of Alaska and the absence of any entertainment." The promise of making a significant sum of money, compared to their home country wages, in a short amount of time, propels young workers to accept these positions. It also allows seafood processing plants to increase their profits by employing a transient, subservient, pool of workers.

Slime lines are located along coasts throughout the world. They are an important part of the production of global seafood commodities, derived from both fisheries and aquaculture operations. This fast-growing commodity sector attempts to reduce labor costs through the employment of children and migrants. Global wage inequalities help maximize profits and facilitate the unequal economic and ecological exchange throughout the world-economy. Migrant labor is also exploited within the processing plants in the global North to further enhance the accumulation process. This commodity sector, given the growth imperative of capitalism, continues to intensify its operations, resulting in the depletion of fish stock throughout the world.

Conclusion

The ecological transformation of marine systems is intimately connected to the structure of the global capitalist economy and the international division of labor. The tragedy of the commodity approach helps reveal how the drive for constant accumulation of capital has influenced the organization of fishing and processing plants and contributed to the expropriation of labor and nature. It emphasizes the unity of social production and nature, and it illuminates distinct contradictions of capitalist commodity production. Nature is perceived as a free gift to capital and does not enter into capitalist accounting. Capital attempts to reduce costs to maximize profits. In regard to some fishing operations in Southeast Asia, cost reduction has involved the exploitation of sea slaves. These operations supply fish for the global market, with the largest shares flowing to Europe and the United States. But due to the overexploitation of fish stocks, driven by efforts to provide seafood for the global North, and declining target fish populations, they often capture trash fish. These fish are used to produce feed inputs, such as fishmeal and fish oil, in aquaculture operations. In Thailand and other Southeast Asian countries, shrimp aquaculture is an important food industry. The majority of shrimp consumed in the global North are produced in aquaculture facilities. These facilities are also well known for the environmental degradation they cause, such as the destruction of crucial coastal ecosystems, especially mangrove forests. Trash fish are also used to produce fishmeal for animal feed, including the production of pet food.

Periphery nations export the bounty of the oceans to the global North and marine populations are depleted. In seafood processing plants, child labor and migrant labor are commonly used, in order to suppress labor costs. This work tends to be hazardous and expose workers to an array of health risks. Global wage inequalities that accompany this international division of labor help ensure that accumulation of capital takes place along the commodity chain within the seafood sector, while deepening social inequalities between classes. Within the global capitalist system, unequal economic exchange and unequal ecological exchange are intimately intertwined.

Capital receives more value in labor power and ecological wealth and services for less. The appropriation of productive environmental space, whether land or ocean, and labor power is intertwined with depleted marine ecosystems. The expropriation of sea life within these ecological systems exacerbates the ongoing exploitation of labor by creating new socioecological dynamics where extracting the free gifts of nature becomes increasingly difficult. The fishing sector in the global South is continually forced to further exploit regional fisheries, as global markets persistently reinforce the unequal relationships within the global hierarchy of nations. As seafood production shifts with the ongoing growth of aquaculture, the depletion of target fish, and the expansion in the production of fishmeal and fish oil, the relationships that connect slave labor, slime lines, and marine degradation become more embedded within a system predicated on the constant accumulation of capital that creates global social and ecological inequalities. The tragedy of the commodity is revealed in the deterioration of marine ecosystems and the human labor on which global capital thrives.

References

Amin, Samir. 1976. *Unequal Development*. New York: Monthly Review Press.

———. 2010. *The Law of Worldwide Value*. New York: Monthly Review Press.

Asia Foundation and International Labor Organization. 2015. "Migrant and Child Labor in Thailand's Shrimp and Other Seafood Supply Chains: Labor Conditions and the Decision to Study or Work." Bangkok, Thailand: Asia Foundation and International Labor Organization.

Bales, Kevin. 2016. *Blood and Earth*. New York: Spiegl & Grau.

Baran, Paul and Paul Sweezy. 1966. *Monopoly Capital*. New York: Monthly Review Press.

Bunker, Stephen. 1984. "Modes of Extraction, Unequal Exchange, and the Progressive Underdevelopment of an Extreme Periphery." *American Journal of Sociology* 89:1017–1064.

Bunker, Stephen and Paul Ciccantell. 2005. *Globalization and the Race for Resources*. Baltimore, MD: Johns Hopkins University Press.

Burkett, Paul. 1999. *Marx and Nature*. New York: St. Martin's Press.

Chase-Dunn, Christopher. 1998. *Global Formation*. Lanham, MD: Rowman & Littlefield.

Clausen, Rebecca and Brett Clark. 2005. "The Metabolic Rift and Marine Ecology." *Organization & Environment* 18(4):422–444.

Cox, Susan Jane Buck. 1985. "No Tragedy on the Commons." *Environmental Ethics* 7(1):49–61.

Dickens, Peter. 2004. *Society and Nature*. Cambridge, UK: Polity.

Emmanuel, Arghiri. 1972. *Unequal Exchange*. New York: Monthly Review Press.

Environmental Justice Fund. 2014. "Slavery at Sea: The Continued Plight of Trafficked Migrants in Thailand's Fishing Industry." London: Environmental Justice Fund.

Foster, John Bellamy. 1994. *The Vulnerable Planet*. New York: Monthly Review Press.

———. 2000. *Marx's Ecology*. New York: Monthly Review Press.

Foster, John Bellamy, Brett Clark, and Richard York. 2010. *The Ecological Rift: Capitalism's War on the Earth*. New York: Monthly Review Press.

Foster, John Bellamy and Hannah Holleman. 2014. "The Theory of Unequal Ecological Exchange." *Journal of Peasant Studies* 41(2):199–233.

Frank, Andre Gunder. 1967. *Capitalism and Underdevelopment in Latin America*. New York: Monthly Review Press.

Frey, R. Scott. 1994. "The International Traffic in Hazardous Wastes." *Journal of Environmental Systems* 23:165–177.

———. 2015. "Breaking Ships in the World-System: An Analysis of Two Ship Breaking Capitals, Alang-Sosiya India and Chittagong, Bangladesh." *Journal of World-Systems Research* 21:25–49.

Galeano, Eduardo. 1997. *Open Veins of Latin America*. New York: Monthly Review Press.

Grivan, Norman. 1976. *Corporate Imperialism*. New York: Monthly Review Press.

Gordon, H. Scott. 1954. "The Economic Theory of a Common-Property Resource." *Journal of Political Economy* 62(2):124–142.

Guevara, Ernesto C. 1997. *Che Guevara Reader*. Melbourne: Ocean Press.

Halpern, Benjamin S., Shaun Walbridge, Kimberly A. Selkoe, Carrie V. Kappel, Fiorenza Micheli, Caterina D'Agrosa, John F. Bruno, Kenneth S. Casey, Colin Ebert, and Helen E. Fox. 2008. "A Global Map of Human Impact on Marine Ecosystems." *Science* 319:948–952.

Hardin, Garrett. 1968. "The Tragedy of the Commons." *Science* 162:1243–1248.

Heilbroner, Robert. 1985. *The Nature and Logic of Capital*. New York: W.W. Norton.

Hornborg, Alf. 1998. "Towards an Ecological Theory of Unequal Exchange." *Ecological Economics* 25:127–136.

———. 2003. "The Unequal Exchange of Time and Space." *Journal of Ecological Anthropology* 7:4–10.

———. 2011. *Global Ecology and Unequal Exchange*. London: Routledge.

International Labor Organization. 2013. *Caught at Sea: Forced Labour and Trafficking in Fisheries*. Geneva, Switzerland: International Labor Organization.

———. 2015. "International Expert Meeting on Labour Exploitation in the Fishing Sector in the Atlantic Region." Fundamental Principles and Rights at Work Branch, Geneva, Switzerland: International Labor Organization.

Irvine, Dean, Saima Mohsin, and Kocha Olarn. 2015. "Seafood from Slavery: Can Thailand Tackle the Crisis in its Fishing Industry?" CNN, May 17. Available at http://www.cnn.com/2015/05/11/asia/freedom-project-thailand-fishing-slave-ships/. Accessed December 15, 2015.

Jeebhay, M. F., T. G. Robbins, and A. L. Lopata. 2004. "World at Work: Fish Processing Workers." *Occupational and Environmental Medicine* 61:471–474.

Jorgenson, Andrew K. 2006. "Unequal Ecological Exchange and Environmental Degradation." *Rural Sociology* 71:685–712.

Jorgenson, Andrew K., Kelly Austin, and Christopher Dick. 2009. "Ecologically Unequal Exchange and the Resource Consumption/Environmental Degradation Paradox." *International Journal of Comparative Sociology* 50:263–284.

Jorgenson, Andrew K. and Brett Clark. 2009. "The Economy, Military, and Ecologically Unequal Exchange Relationships in Comparative Perspective." *Social Problems* 56(4):621–646.

———. 2011. "Societies Consuming Nature." *Social Science Research* 40:226–244.

Kammer, Jerry. 2011. "Cheap Labor as Cultural Exchange: The $100 Million Summer Work Travel Industry." Washington, DC: Center for Immigration Studies.

Lawrence, Kirk S. 2009. "The Thermodynamics of Unequal Exchange." *International Journal of Comparative Sociology* 50(3–4):335–359.

Longo, Stefano B., Brett Clark, Thomas E. Shriver, and Rebecca Clausen. 2016. "Sustainability and Environmental Sociology: Putting the Economy in its Place and Moving Toward an Integrative Socio-Ecology." *Sustainability* 8(5):437. https://doi.org/10.3390/su8050437.

Longo, Stefano B., Rebecca Clausen, and Brett Clark. 2015. *The Tragedy of the Commodity: Oceans, Fisheries, and Aquaculture*. New Brunswick, NJ: Rutgers University Press.

Magdoff, Harry. 1978. *Imperialism*. New York: Monthly Review Press.

Marx, Karl. 1976. *Capital, Volume 1*. New York: Vintage.

———. 1991. *Capital, Volume 3*. London: Penguin.

———. 1993. *Grundrisse*. London: Penguin.

Mill, John Stuart. 1877. *Essays on Some Unsettled Questions of Political Economy*. London: Longmans, Green, and Co.

Millennium Ecosystem Assessment. 2005. *Ecosystems and Human Well-being: Synthesis*. Washington, DC: Island Press.
Muszynski, Alicija. 1996. *Cheap Wage Labour, Race, and Gender in the Fisheries of British Columbia*. Montreal, Canada: McGill-Queen's University Press.
O'Connor, James. 1998. *Natural Causes*. New York: Guilford Press.
Pauly, Daniel. 2009. "Beyond Duplicity and Ignorance in Global Fisheries." *Scientia Marina* 73(2):215–224.
Radke, August and Barbara Radke. 2002. *Pacific American Fisheries, Inc: History of a Washington State Salmon Packing Company, 1890–1996*. Jefferson, NC: McFarland & Company.
Ricardo, David. 1951. *On the Principles of Political Economy and Taxation*. Cambridge: Cambridge University Press.
Rice, James. 2007. "Ecological Unequal Exchange." *Social Forces* 85:1369–1392.
———. 2008. "Material Consumption and Social Well-being within the Periphery of the World-Economy." *Social Science Research* 37:1292–1309.
———. 2009. "The Transnational Organization of Production and Uneven Environmental Degradation and Change in the World-Economy." *International Journal of Comparative Sociology* 50(3–4):215–236.
Roberts, Callum. 2012. *The Ocean of Life*. New York: Penguin.
Roberts, J. Timmons and Bradley Parks. 2007. *A Climate of Injustice*. Cambridge, MA: MIT Press.
Roy, Patricia. 1989. *A White Man's Province: British Columbia, Politicians, and Chinese and Japanese Immigrants, 1885–1914*. Vancouver, Canada: University of British Columbia Press.
Sanderson, Matthew R. and R. Scott Frey. 2014. "From Desert to Breadbasket to Desert Again? A Metabolic Rift in the High Plains Aquifer." *Journal of Political Ecology* 21:516–532.
Schaefer, Milner B. 1957. "Some Considerations of Population Dynamics and Economics in Relation to the Management of the Commercial Marine Fisheries." *Journal of the Fisheries Board of Canada* 14(5):669–681.
Schnaiberg, Allan. 1980. *The Environment*. New York: Oxford University Press.
Scott, Anthony. 1955. "The Fishery." *Journal of Political Economy* 63(2):116–124.
Shandra, John, Christopher Leckband, Laura McKinney, and Bruce London. 2009a. "Ecologically Unequal Exchange, World Polity, and Biodiversity Loss." *International Journal of Comparative Sociology* 50:285–310.
Shandra, John, Eran Shor, and Bruce London. 2009b. "World Polity, Unequal Exchange, and Organic Water Pollution." *Human Ecology Review* 16:51–64.
Smith, Adam. 1937. *The Wealth of Nations*. New York: Modern Library.

Smith, John. 2016. *Imperialism in the Twenty-First Century*. New York: Monthly Review Press.

Srinivasana, U. Thara, Susan Carey, Eric Hallstein, Paul Higgins, Amber Kerr, Laura Koteen, Adam Smith, Reg Watson, John Harte, and Richard Norgaard. 2008. "The Debt of Nations and the Distribution of Ecological Impacts from Human Activities." *Proceedings of the National Academy of Sciences* 105(5):1768–1773.

Stretesky, Paul and Michael Lynch. 2009. "A Cross-National Study of the Association Between Per Capita Carbon Dioxide Emissions and Exports to the United States." *Social Science Research* 38:239–250.

Sweezy, Paul. 2004. "Capitalism and the Environment." *Monthly Review* 56(5):86–93.

United Nations. 2011. "FishStatJ: Universal Software for Fishery Statistical Time Series." Rome, Italy: Food and Agriculture Organization.

———. 2014. "State of World Fisheries and Aquaculture, 2010." Rome, Italy: Food and Agriculture Organization.

United Nations Inter-Agency Project on Human Trafficking (UNIAP). 2009. "SIREN Case Analysis Exploitation of Cambodian Men at Sea." Bangkok, Thailand: United Nations.

U.S. Bureau of Labor Statistics. 2012. "Labor Force Characteristics by Race and Ethnicity, 2012." Available at http://www.bls.gov/cps/cpsrace2012.pdf. Accessed December 28, 2015.

U.S. Department of State. 2015. "Trafficking in Persons Report." U.S. Department of State Publication. Office of Undersecretary for Civilian Security, Democracy, and Human Right. Washington, DC: US Department of State.

Urbina, Ian. 2015a. "Tricked and Indebted on Land, Abused or Abandoned at Sea." *New York Times*. Available at http://www.nytimes.com/2015/11/09/world/asia/philippines-fishing-ships-illegal-manning-agencies.html?_r=0. Accessed December 10, 2015.

———. 2015b "'Sea Slaves:' The Human Misery that Feeds Pets and Livestock." *New York Times*. Available at http://www.nytimes.com/2015/07/27/world/outlaw-ocean-thailand-fishing-sea-slaves-pets.html. Accessed December 10, 2015.

9

History Matters: Contingency in the Creation of Ecologically Unequal Exchange

Shellen Xiao Wu

In her new book, *The Mushroom at the End of the World*, Anna Tsing (2015) traces the matsutake mushroom from the post-industrial forests of the Pacific Northwest, the scarred and exploited woodlands in Yunnan in Southeast China, to mushroom suppliers and middlemen, and finally, as the beautifully presented product of a culture of gift-giving in Japan. Matsutake mushrooms grow particularly well in a relationship of interdependency with second-growth pine forests, which proliferate in high traffic, human-disturbed woodlands. The pines depend on the mushrooms to break down soil in nutrient-poor terrains and areas with little to no undergrowth. In turn, the mushrooms depend on the pines, which grow better than other trees in scarred landscapes cleared by fire and industrial forestry. The ecology of industrial ruin paradoxically fosters the rise of profitable mushroom trade. The matsutake commodity chain exposes the global linkages of capitalism but also its points of discontinuities. The interstices and blind spots of capitalist accumulation and commodification allow former Hmong and Cambodian refugees, Vietnam veterans,

S. X. Wu (✉)
Department of History, University of Tennessee, Knoxville, TN, USA
e-mail: swu5@utk.ed

R. S. Frey et al. (eds.), *Ecologically Unequal Exchange*,
https://doi.org/10.1007/978-3-319-89740-0_9

and Japanese American pickers to each find a space of freedom outside the confines of modern capitalism. Tsing makes clear the importance of history, contingency, and interdependence in the creation of mushroom landscapes but also more broadly in our twenty-first-century reassessment of capitalism and science.

We live in a human-disturbed world, where the tendrils of capitalism have extended to the furthest reaches of the globe. Yet, Tsing's work exposes the unquantifiable zones outside commodity chains and capitalist markets. These regions where salvage accumulation takes place also happen to fall into the gray zones of academic disciplines, somewhere between science, social sciences, and history. For Tsing, if "progress depended on conquering an infinitely rich nature through alienation and scalability," then matsutake uncovers the non-scalability of the world (Tsing 2015:135; see also Tsing 2012). In these areas, the ecology is the result of human development and the people who participate in salvage accumulation the human debris of history, of wars, industrialization, and migration. The mushroom pickers cherish the freedom of their pastime. Yet, the sale of their pickings in "open ticket" ("an informal mushroom buying market in the woods") connects them back to trading spots, markets, commodity chains, and ultimately global capitalism (Tsing 2015:74–83). In other words, their livelihoods and freedom from the workplace hierarchy, much like the interdependent relationship between matsutake and the ecology of industrial ruin, exist both because of and despite unequal exchange.

On the surface coal seems far removed from the matsutake as a commodity and as the product of the often exploitative relationship between capital and labor. Coal has played an outsized role in the world-system literature as the commodity that fueled industrialization and ultimately the rise of imperialism in the nineteenth century. Each coal deposit has unique characteristics depending on the geology of its location, its own history as it were. These characteristics arising from the chemical makeup of individual deposits directly affected the development of technologies that used coal, from steamship and train engines to stoves and lamps (Shulman 2015). In turn, these technological innovations drove both global and local demand for coal. The presence of coal deposits did not necessarily lead to industrialization. An entire economic literature has sprouted on the so-called resource curse to describe the phenomenon of resource-rich countries which nevertheless develop more slowly than other countries with fewer natural resources (Sachs and Warner 2001). Yet on closer examination, what is chalked up to the "curse" is in fact often the result of particular histories in particular places. We can apply the lessons of the matsutake to coal. In both instances, historical contingency led to a chain of events that allowed each to take its place in a global network of capital.

Historical contingency and ecological difference introduce what Tsing (2005) terms "friction" to the study of capitalism, or the unquantifiable result of specific places and contexts. For example, anthracite gained prominence on the American market only after the discovery and exploitation of major deposits in western Pennsylvania and the construction of expensive infrastructure, including a canal system to transport coal from mines in the interior to markets in East Coast big cities like Philadelphia, Baltimore, and New York. Initially, consumers found top-grade anthracite difficult to light and the resulting fire hard to control. Some consumers could not light the coal and dumped shipments on the streets to use as paving stones. Mining companies needed engineers and inventors to adjust the design of various technologies from engines to stoves in order to create the market for anthracite (Jones 2010, 2014). The expensive process of creating demand financially ruined more than a few of the early pioneers in the American coal industry. To speak of coal as a uniform commodity obscures the distinctive historical context for the use of resources and patterns of consumption. In the United States, this history explains the dominance of the Northeast and the Mid-Atlantic region in nineteenth century industrialization. Inland regions that provided the coal and oil bore the environmental costs for the rapidly growing energy use of major East Coast cities (Jones 2014:66–70, 86). At the same time, regions outside these networks, including the American South, quickly fell behind in energy use. The ecologically unequal exchange of the American energy market resulted from the accident of geology and was subsequently amplified by the uneven development of infrastructure.

The American case study was hardly unique. In other parts of the world, ecologically unequal relationships evolved out of specific environmental constraints and historical contingency. By focusing on a case study with which I am most familiar—coal and mining in China—I aim to bring a historical perspective to the discussions of this volume on ecologically unequal exchange. While the world-systems literature provides a useful framework for understanding global connections, Bunker (1985:38, Bunker, page 27, this volume) pointed out the limitations of generalized theory, arguing that "theories which assign explanatory primacy either to global or to regional systems ignore historical processes, continuities, and dynamics in other systems." Energy extraction and use developed in China on a very different trajectory than in Europe or the Amazon Basin in Brazil that Bunker focused on in his fieldwork. This difference resulted from unique aspects of Chinese geography and geology, as well as China's response to imperialism from the late nineteenth century. The country's sheer size and competing interests between Western powers resulted in China's escape from full colonization. As a result, while ecologically unequal exchange is present *within* China, the country as a whole did not fall into the development trap of other third world countries. The so called "resource curse" did not strike China, despite considerable Western interest in its coal reserves in the nineteenth century.

The Ecological Limits of Growth

Over the course of the nineteenth century fossil fuels replaced wood as the dominant form of energy used in human society. The transition in energy regime that began as a result of geographically specific environmental constraints led to a series of technological innovations, which in turn informed a growing divergence between industrialized countries and the rest of the world. Historians, economists, and other social scientists from Max Weber and Werner Sombart at the start of the twentieth century to John Ulric Nef, Fred Cottrell, Edward Anthony Wrigley, and Rolf Peter Sieferle into the twenty-first pinpointed coal as a key factor in the creation of a modern capitalist world order (Berman 1982; Cottrell 1955; Nef 1966; Schivelbusch 1986; Sieferle 2001; Smil 1994; Sombart 1902; Weber 1904/1958; Wrigley 1988; Sugihara 2003). The importance of coal supplies depended on their location, particularly proximity to transportation, as well as suitability for industrial use.

Environmentally, both Europe and China entered into a period of intensifying use of natural resources in the sixteenth century. Deforestation in Europe meant that the large logs used to build ships had to travel longer distances to reach port cities. England already switched to the use of coal in the sixteenth century for heating and metallurgy. Similarly, major centers of the porcelain industry in China, in Jingdezhen in Jiangxi province, for instance, used coal for kilns as the surrounding countryside became denuded of large forests. In order to satisfy growing population demands, women and children turned to piecework and home weaving to produce extra income. Increasing efficiency in human labor nevertheless failed to fully counter ecological constraints: the limited supplies of commodities, such as cotton for homespun, and of energy sources (Elvin 1972).

At this critical juncture, the opening of the New World opened up the logjam for Europe. In Jason Moore's (2015:85, 144) recent work, *Capitalism in the Web of Life*, he places the impetus for capital accumulation in the progressive opening of various forms of frontiers—agrarian frontiers as well as the vertical frontiers of coal and oil. For Moore (2015:132), modernity's energy revolution dates back to increases of coal production starting in the 1530s in England. He traces capitalism's current crisis to the end of Cheap Nature and the seemingly endless geographical expansion and technological innovations of the past 500 years, which stimulated and enabled the accumulation of capital by the progressive appropriation of new and unpaid sources of work. Moore adds to the discussion a much-needed spatial component of analysis. From the 1970s, the expansion of capitalism encountered the exhaustion of natural resources and new frontiers, both on Earth's surface and vertically in the subterranean realms. In the last few years, deep sea drilling and fracking, as well as geopolitical realignment of new resource frontiers in Brazil and Russia, have led OPEC to increase production to maintain market share, resulting in the virtual collapse of oil prices by late 2015. Yet, paradoxically, low energy prices have failed to stimulate global economies and have instead appeared to portend further global turmoil in the markets and deepening recession in developing countries dependent on China's seemingly insatiable energy demands. Moore's work updates both the work of classical theorists like Werner Sombart and Max Weber, as well as world-systems theorists since Wallerstein (1974–2011), including Andre Gunder Frank (1998) and Giovanni Arrighi (2007, 2010). He connects this long-standing debate with growing concerns about the ecological costs of industrialization, costs that thus far have been largely hidden or at least underestimated in the discussion about nature and capital.

Moore (2015) did not focus on China's role in the creation and decline of Cheap Nature. Frank (1998) and Arrighi (2007, 2010), on the other hand, both recognized the temporary ascendance of Europe as the center of the modern world order from circa 1450 to the long nineteenth century, the decline of the American core since the 1970s, and since the 1990s, China's resumption of its place at the core of a Sino-centric world order (Arrighi 2007; Frank 1998). Both scholars used Marxism, capitalism, and environmental histories of energy to modify Wallerstein's (1974–2011) classical formulation of world-system theory. In his later works Arrighi located East Asia's lack of overseas empires as the point of departure for the great divergence. Arrighi's (2007) *Adam Smith in Beijing* and subsequently Moore's (2015) Great Frontier both employed historical arguments to explain capitalism's global spread, as well as its limitations. With his background in geography, Moore is acutely attuned to the spatial dimension of capitalism's spread. The opening of new frontiers through the nineteenth century resulted in the effective subjugation and removal of American Indians to arid reservations—until even these areas, with the discovery of oil and other natural resources—became contested zones. Oklahoma and the Dakotas, for example, became desirable places only after the discovery of fossil fuel deposits underground. Cheap Nature benefited from the free or undercompensated labor of colonized and conquered peoples, such as the American Indians or the Indians in Asia, whose wealth and agricultural output paid for the British Empire. In return the Indian Subcontinent experienced nearly zero economic growth from 1820 to the 1870s and the effective destruction of a thriving local textile industry through their incorporation into the British commodities/industrial market. Cheap Nature was thus bound to the dispossession and unequal relationship of the colonial system. Indigenous peoples and women and children enabled Cheap Nature because they failed to receive fair compensation for their lands, resources, and work. By this logic, modern capitalism both created and maintained ecologically unequal exchange, encompassing even Anna Tsing's (2015) mushroom pickers in their salvage accumulation.

China did not have the good fortune of opening the New World frontiers to break through its ecological constraints. Nevertheless, coal played an important role in China's entry into the modern world order. The importance of coal to industrialization was only recognized in retrospect. China has a long history in the mining of various metals, stones, and coal. Archaeological excavations have uncovered mines dating from the Warring States period (475–221 BCE) that already used sophisticated timbering methods (Golas 1999:11–12). Between 750 and 1000 CE the production of pig iron grew sixfold, relying for fuel on massive quantities of coal (Hartwell 1967:104). The environmental historian Mark Elvin (1972) first formulated the "high-level equilibrium trap" to explain why industrialization started in England rather than in China, despite more advanced technologies and agricultural production in the latter already in the fourteenth century. Elvin argued that the greater efficiencies of the domestic trade network, in addition to China's more advanced technological development, including in areas like textiles and mining, led to incremental increases in productivity and eventually stagnation. The fundamental idea behind the "high-level equilibrium" had been around since the 1930s, when Western missionaries and developmental experts first attempted to enact rural reform in China (Zanasi 2013). But whereas in the 1930s Western experts formulated their ideas based on very limited knowledge of Chinese environmental conditions, Elvin progressively refined his argument through a *longue durée* analysis of Chinese environmental history. Using a variety of sources, including poetry and literati writings, Elvin uncovered a utilitarian view of nature in China that dated to antiquity. For Elvin, extensive references from Chinese literature displayed a pattern of extensive exploitation of natural resources in China long before the Mao era and its calls for the conquest of nature through environmentally disastrous large-scale engineering plans (Elvin 2004). Although today we are most familiar with the man-made disasters of the 1960s and 1970s, including structurally unsound dams, strip mining, and massive Soviet-style industrial works, environmental degradation in China had begun long before the modern era (Shapiro 2001).

Subsequently, other Chinese environmental historians have come to similar conclusions. Robert Marks (1998, 2011) has shown how in order to feed a rapidly growing population, particularly in the outwardly prosperous seventeenth and eighteenth centuries, the late imperial Chinese state and bureaucracy turned to intensive exploitation of nature as a solution to exponentially escalating demands for resources. Mid-eighteenth century observers noticed the collapse of copper-bearing hills after a period of intensive mining, as well as the depletion of soil fertility on newly opened frontier lands (Marks 2011:220). In the late eighteenth century, the official Hong Liangqi warned Qing officials of impending crisis from an unchecked rise in population. Much like the English author Thomas Malthus, Hong argued that "The amount of [available land and housing] has only doubled, or at the most, increased three to five times, while the population has grown ten to twenty times.... the resources with which Heaven-and-earth nourish the people are finite" (Marks 2011:221). These were not isolated instances. Chinese officials, however, did not look outwards for solutions to the issue of environmental depletion. Instead, eighteenth and nineteenth century officials subscribed to a statecraft school of governance, which emphasized the cultivation of practical industries, such as the silk farming or mining, to foster economic growth (Rowe 2001).

Despite a devoted coterie of provincial officials who worked to foster growth and provide practical solutions for problems in the areas under their jurisdiction, there is considerable evidence that the rapid growth of the Chinese population created increasing ecological constraints by the late eighteenth century. By examining both Chinese and Jesuit discussion of agricultural practices in the late imperial period, Elvin (2004) argues that, based on contemporary accounts, most of China's areas of intensive agricultural development did not practice fallowing, a prevalent agricultural practice in Europe. Population pressures created the need for fields to produce yearly, and restoration of soil fertility relied on intensive application of fertilizers. Given the paucity of large draft animals in most parts of the country, restoration of soil fertility relied on human excrement as the primary fertilizer. Farmers applied fertilizer not just to the most important crops, but every plant, multiple times during the growth cycle. Jesuits traveling in parts of the country outside of the capital in Beijing and other major cities noted arid and desolate mountainscapes. The dense population meant that exhausted soils nevertheless needed to produce multiple crops per year (Elvin 2004:466).

More recent works in environmental history similarly point to the exhaustion of Chinese soil fertility in significant parts of the country by the nineteenth century. John Richard's (2003) comparative study of frontiers contrasted the success of forestry efforts in Japan to the escalating costs of hydraulic maintenance in China. In some domains of Japan, agricultural writers advocated the use of fishmeal as a source of fertilizer for agriculture (Arch 2015; Richards 2003). On the other hand, by the mid-nineteenth century in China even the intensive application of fertilizer made little difference as ecological damage forced farmers off the land. Impoverished Chinese peasants resorted to finding employment in the international coolie trade. Conditions in South American plantations that employed coolie labor were brutal and akin to slavery. Few of those recruited for the coolie trade returned home alive. Yet, those who joined the coolie trade may not have had a choice. Gregory Cushman's (2013) examination of the Pacific guano trade shows how ecological collapse and rural dispossession in China led to the trans-Pacific coolie trade in the nineteenth century. Countless coolies committed suicide rather than continue to work in inhumane conditions (Clark and Foster 2009; Cushman 2013). Historically contingent environmental conditions resulting from rapid population growth coupled with bureaucratic stagnation helped to create this condition of ecologically unequal exchange. The export of coolie labor suggests that rather than turning to coal for an energy breakthrough, China released its excess population pressures by other means through the growing linkages of global commodity chains.

The question of coal and China's divergent path of development from Europe returned to prominence with Kenneth Pomeranz's (2002) publication of *The Great Divergence*. Pomeranz revisited some of the issues central to the "high-level equilibrium trap" thesis, but also expanded his lens to the resource frontiers. In both Europe and China, he argued, technological innovations and piecework labor contributed to incremental improvements in productivity. He made the strongest case for the argument that what made the difference for Europe to make a giant leap in progress, leading to industrialization, was the opening of the New World

Frontiers—the Cheap Nature of Jason Moore's (2015) telling. Environmental conditions played an essential role in long-term shifts in the economic and commodity ties between different world regions. Parthasarathi (2011) has similarly argued for environmental differences as the cause of the great divergence. Both Parthasarathi (2011) and Pomeranz (2002) located the divergent fates of China and Europe around the late eighteenth century, before which Asian populations enjoyed living standards on par if not above their European counterparts. Parthasarathi has further argued for considerably different environmental constraints between China and India in the seventeenth and eighteenth centuries. He pointed out that the Indian subcontinent did not have the same ecological constraints and shortage of wood as England (Parthasarathi 2011:13). In the seventeenth and eighteenth centuries, significant portions of India remained densely wooded, unlike the denuded landscape of England by that time, which spurred the transition to coal.

China, however, did have deforestation problems, as well as significant coal deposits. Pomeranz (2000) pointed to the geographical disadvantage of coal supplies in China, which were clustered in the Northwest. Remote coal mines had no incentive to increase production because transportation problems prevented them from supplying the fuel needs of large cities (Pomeranz 2000:63). Pomeranz further argued that geological differences between English and Chinese mines acted as a further disincentive for Chinese innovation; Chinese coalmines for the most part did not face significant problems with water accumulation requiring the use of mechanical pumps, the initial use for steam engines. The original Newcomen steam engines were so inefficient that outside of collieries the cost of fuel made their use prohibitively expensive and impractical. Ventilation, rather than ways to remove water as in England, was the chief technical problem in Chinese mines. Pomeranz (2000:66–67) also cited the distance of major coal supplies from the wealthy Yangzi Delta region as a factor in limiting the transmission and advancement of technological expertise. Parthasarathi (2011:162–164) agreed with Pomeranz on the importance of geography and environment but differed on the question of coal, arguing that coastal regions did in fact have access to coal. Both works explore the key question of why the availability of coal did not necessarily lead to industrialization and whether there are other factors in addition to the conditions listed above.

It would not be until the twentieth century before Chinese writers recognized, as did Max Weber (1904/1958), the connection between fossil fuels and the capitalistic world economic system. By the early twentieth century, discussion of coal lost the specificity of local needs and instead elevated it to the essential fuel of industrialization. I have argued elsewhere that the Chinese transition in energy regime and adoption of coal as the fuel of industrialization happened in discourse and in law long before official statistics reflected this change (Wu 2014, 2015). These changes were a historically and place-specific response to the arrival of Western science and technologies. Coal went from a useful mineral to a fuel for industrialization by the end of the nineteenth century. Geographical difference and historical contingency both contributed to a distinctively Chinese trajectory of industrialization. The historical argument grounds theoretical discussions of capitalism, world-systems, and energy regimes in specific contexts and provides insight to the non-scalable aspects of the coal question. Coal was widely used for heating purposes in north China and in other parts of the country with coal supplies. Farmers mined coal in the winter months. Most output did not travel very far from the mines because while officials saw coal as important for people's livelihoods, it was not considered a valuable commodity. In contrast, silver and copper used in East Asian currencies underpinned inter-Asia trade and made up significant proportions of Chinese and Japanese exports and imports (von Glahn 1996:209–229). For these metals, the Qing court oversaw an elaborate transport network that moved silver and copper from mines in the Southwest to mints in the capital, over steep and treacherous roads.

The sheer size and geographical diversity of the Chinese mainland, as well as a waning but still powerful Qing state kept imperialist interests at bay until fairly late in the Great Powers land grab in the nineteenth century. In the years between the Sino-Japanese War (1894–1895) and the collapse of the Qing dynasty in 1911, however, Western powers gained significant mining and railroad rights in provinces across China. Britain, France, Germany, and the other Western powers attempted to create a resource frontier in China as they had earlier done so with guano in Peru, diamond and precious metals in Africa, and coal in fueling depots around the world. To counter foreign demands for mining rights in the interior, both the central government and provincial elites adopted the separation of surface and mining rights and sought to increase state control over minerals, in some provinces establishing outright monopolies over all mining activities. Despite the declining fortunes of the ruling Qing regime in the nineteenth century, China was never fully colonized and as a result, officials and reformers had considerable leeway to impose their own set of laws and regulations on the mineral industry (Wu 2012).

The failure of Western encroachment on Chinese mineral rights was certainly not due to the lack of interest (Wu 2014). In 1885, the German geographer Ferdinand Richthofen published a folio of maps of China as a supplement to his China volumes based on a series of seven expeditions he conducted in China from 1868 to 1872. Prior to Richthofen's (1903) discoveries, the state of Pennsylvania contained the world's largest known coal-fields by area; merely the province of Shanxi, wrote Richthofen, would dwarf Pennsylvania's coal deposits. For centuries China had produced exquisite silks, porcelain, and exported tea, but Richthofen's maps suggested its true treasures for a power hungry, modernizing Europe—what lay beneath the surface of the land. In addition to the maps, during his time in China Richthofen wrote a series of detailed letters during 1870–1872 to the British controlled Shanghai Chamber of Commerce on the mineral deposits in the Chinese interior. Richthofen (1903) wrote a total of 11 letters from 1870 to 1872 and discussed in these letters Hunan, Hubei, Henan, Shanxi, Zhejiang, Gansu, the area around Nanjing, Sichuan, Zhili, and Mongolia. He devoted a section in each letter to the topography of the regions he traveled through but focused on the valuable exports of each province, and chief among these the mineral products.

In Shanxi Province, Richthofen (1903:43) described the undulating highlands in the southern regions of the province, with a thick layer of loess covering the ground and intersected by deep watercourses:

> It will be seen that Shansi is one of the most remarkable coal and iron regions in the world; and some of the details which I will give will make it patent that the world, at the present rate of consumption of coal, could be supplied for thousands of years from Shansi alone. Professor Dana, in comparing the proportions in which, in different countries, the area of the coal

> land is to the total area, says: - "The State of Pennsylvania leads the world, its area of 43,960 square miles embracing 20,000 of coal land." It is very probable that, on closer examination, the province of Shansi in China, with an area of about 55,000 square miles, will take the palm from Pennsylvania, by a considerably more favorable proportion. But this is not yet all the advantages on the side of the Chinese coal fields. Another is afforded in the ease and cheapness with which coal can be extracted on a large scale.
>
> On the other hand, the whole of this great coal and iron region labours under two great disadvantages. Firstly, it is situated a distance away from the coast, and from rivers that are fit for other navigation than by small Chinese boats; and secondly, the whole of the coal formation rests, as it were, on a platform raised a few thousand feet above the adjoining plain. Its steep descent to the latter will not form an obstacle, but at least offers great difficulties, to the construction of a railroad, which will be the only means of ever bringing to account the mineral wealth of Shansi.

To provide a better conception of the massive extent of the Shanxi coal deposits, Richthofen cited the American geologist James Dwight Dana's system of ranking countries and their coal potentials. Dana had measured a country's coal lands as a ratio to the total area. The resulting measurement presented a new way of looking at a country's wealth and industrial potential.

Richthofen's letters became a turning point in foreign interest in Chinese mineral deposits, although the floodgates of European demands for mining concessions did not open until after China's humiliating loss in the Sino-Japanese War (1894–1895). Saddled with a considerable indemnity payment for the war, the fiscal situation of the central government deteriorated rapidly. For European powers and individual fortune hunters, the weakened Qing government appeared an easy target. As a reaction to the German acquisition of Jiaozhou in Shandong province in 1898, where the Russians had also shown interest, the Russians demanded Port Arthur on the Liaodong peninsula as compensation. Soon thereafter the British demanded Weihaiwei on the other side of the peninsula to counter Russian expansion. On May 21, 1898, the British company Peking Syndicate signed a contract with the Shanxi provincial government for the exclusive rights to develop extensive coal and iron rights in several areas in the province, along with the right to drill for petroleum throughout the province. The company's agent in China, an Italian named Luzatti, shortly also concluded a similar contract with Henan province. Yet, on the brink of triumph, the tide turned against foreign concession hunters. The year 1905 became a watershed moment in the provincial movement to reclaim mines.

Instead of folding to foreign demand for mining rights, the Qing state promulgated a formal set of mining laws in 1902. Two years later in April 1904, a second set of provisional regulations with 38 articles provided the legal basis for mining enterprise in China until the Throne approved a more permanent code in 1907. With each revision, the regulations increased in number, from 19 clauses to 78 in 1907, plus 73 supplementary regulations. The increasing length of the regulations reflected growing awareness of foreign mining laws (Wu 2012). Between 1905 and 1906 a Qing Government Reform Commission went abroad to Europe, the United States, and Japan. Upon the Commission's recommendation, in 1906 the old structure of government was dissolved and replaced by a constitutional monarchy supported by a European-style ministerial system (Horowitz 2003). In the new system, supervision of agricultural, commercial, and industrial matters, including mining related matters were combined under the aegis of the Ministry of Agriculture and Commerce. In the following years, the new Ministry established a Geology Department at Peking University in 1911 and in 1913 opened a Geological Research Institute. In attempting to restrict foreign concession demands, China paradoxically moved closer to Western standards, while also vastly expanding the state's role in mining rights. These regulations served as the model for subsequent revisions even after the Qing collapsed in 1911. In contrast to Latin American countries like Chile and Peru and Middle Eastern states like Egypt, the Qing government and provincial elites recognized the trap of leveraging the country's mineral resources to fund extravagant lifestyles.

At the same time that the central government reformed its mineral policy, provincial elites recognized the threat of foreign investment and control over the mining industry. Even before the central government issued new mining laws, some provincial governments already acted preemptively to extend state control over all mineral rights (Wu 2015:146–147). Between

1905 and 1911, provincial government action and gentry agitation in Zhejian, Fujian, Sichuan, Zhili, Anhui, and Yunnan provinces all attempted to reclaim lost mining rights and ensure the protection of the remaining Chinese mineral deposits. Upon closer examination, in each of these cases, the provincial government paid large sums of money to "reclaim" largely undeveloped potential mines and a few rusting machines. In the short term, Western shell companies, which often seized as much territory as they could but invested little capital in the actual operation of mines, clearly had the better end of the bargain. Over the long term, however, gentry agitation and the nation-wide movement to reclaim mining rights ensured Chinese control over its energy regime (Lee 1977; Wu 2015).

The Global and the Local

Geographical differences and historical contingency have led China to diverge significantly from earlier examples of frontier settlement and economic development, particularly in the Americas, at least until the recent decades. Manifest Destiny and the repeated extension of the line of frontiers effectively diluted the environmental costs of capitalist development in the United States. On the other hand, the smog covering most cities in China serves as visible reminder of the terrible price of wealth and power. For most of the twentieth century industrial works in China have continued to cluster on the eastern coast and major river-ways. The Open Up the West campaign from 1999 has had limited success (Wright 2012). For most of the twentieth century China did not reap the benefits of Cheap Nature. From manufacturing hubs to high tech factories that produce Apple products, economic development has come at an enormous environmental price. Every year when temperatures drop, we see startling images come from smog-covered cities in China's Northeast, of air pollution so thick with haze that it reduces to vague shapes newly built high rises and the few brave souls who venture forth on the streets. In November and December of 2015, right before the Paris Climate talks, air particle levels in the provincial capital of Shenyang in the Northeast reached unprecedented hazardous levels, filling local hospitals with people suffering from respiratory problems. The Chinese population as a whole, rather than a removed minority or a distant colonial possession, has paid the price of industrialization and economic development.

Environmental costs are not evenly distributed. Mining towns the world over are terrible places to live. Soot and smoke cast a pall in these places, whereas the benefits and wealth from industrialization flow to urban centers. After over a century of producing the coal that led to China's industrial rise, the city of Pingxiang in Jiangxi Province is still the backwaters. The longest lines in the city are at the train station—for trains out. The coal city of Fushun in the Northeast is at serious risk of collapsing into a sinkhole from a century of strip mining. This unequal distribution of costs and benefits discourages action, because in the cities people do not see the full extent of the environmental damage at the locations of extraction. However, today the pollution has become terrifyingly visible to those living in the capital and urban hubs (Shapiro 2012). Nor does the pollution stop at national borders. Emissions from China have crossed the Pacific, yet pollution still does not figure into the costs of trade (Wong 2014).

Only in recent years have patterns of mineral exploitation in China begun to resemble the United States. The coal mining industry has undergone major changes in recent decades. The inexorable push outward of China's energy frontiers has continued apace, if anything speeding up from the 1990s. Tim Wright's (2012:28–29) recent work on the Chinese coal industry has shown how coal production over the course of the twentieth century progressively moved from provinces like Shandong, Liaoning, and Hebei, to the frontier areas of Inner Mongolia, Xinjiang, and Shanxi as older deposits were exhausted. The visual presentation of the change in the mining industry displays the progressive geographical spread of the environmental costs of mining to China's Western and peripheral regions (Klinger 2015). These regions, of course, also happen to be occupied by ethnic minorities. The expansion of China's energy frontiers had begun in the 1930s and continues today overseas as Chinese state owned mining and petroleum companies have staked future expansion on investments in Australia, South America, and Africa.

In addition to the country's continuing reliance on coal, lax enforcement of environmental laws has resulted in the heavy metal contamination of significant acreages of arable soil, the despoliation of interior waterways and coastal wetlands, and compromises to the food supply chain (Economy 2010; He 2014). As Elvin (2004) has successfully shown, environmental destruction in China has pre-modern roots. The Mao years and its mantra of socialism and science conquering nature increased the pace of ecological damage, but the rapid economic development of the last three decades has proven the most destructive yet (see also Shapiro 2001, 2012). The growing appetites and spending powers of a rising middle class have only accelerated environmental degradation. At the same time, these environmental issues are also paradoxically proof of Chinese success. These recent developments are possible because of actions officials and elites had taken over a century earlier to forestall Western encroachment on Chinese mining rights. This early intervention prevented China from falling into the category of countries suffering from the "resource curse."

Understanding the historically specific causes of environmental change is essential to how we think about and frame ecologically unequal exchange. There is no arguing that the enormous wealth of the Asian networks of trade drew European interest and the accidental discovery of the New World. In turn, the resource frontiers of the New World allowed successive European powers to break out from marginalized peripheries of an Asia centered world-system. In the twenty-first century, ecologically unequal relations continue in different forms around the world. China has entered into the resource competition and is attempting to establish neo-colonial relationships with resource rich countries in Africa and Latin and South America. In his recent book *The Black Hole of Empire*, Partha Chatterjee (2012:338) examines the way imperial power created and brought together the modern discourse of political, economic, and legal knowledge. The great irony of the twenty-first century and the recent growth in Asian economies is the way that the active intervention of the post-colonial nation state has replicated and extended the practice of imperial power, while at the same time disowning the history of imperialism (Chatterjee 2012:340; Harvey 2005).

Ecologically unequal relations started with the attempt to push the environmental costs of development and capitalism onto others while enjoying its benefits, hence its close relationship with colonialism, imperialism, and the unequal status of the periphery and core. Such examinations, however, require historical context. Beyond numbers and data points, the individual and entangled histories of countries in the global web can help us understand the source of inequality and perhaps, too, provide a point of departure for a solution. The relationship between the global and the local mutually influence each other. Coal may have played a starring role in world-systems analysis, but notions of its scalability needs to be tempered by history and placed into context, including shifting ideas about the relationship between humans and Nature. In the ecologically unequal exchange, history matters.

References

Arch, Jacobina. 2015. "Whale Oil Pesticide: Natural History, Animal Resources, and Agriculture in Early Modern Japan." Pp. 93–112 in *New Perspectives on the History of Life Sciences and Agriculture*, edited by Denise Phillips and Sharon Kingsland. Berlin: Springer.

Arrighi, Giovanni. 2007. *Adam Smith in Beijing: Lineages of the Twenty-First Century*. New York: Verso.

———. 2010. *The Long Twentieth Century: Money, Power and the Origins of Our Times*. New York: Verso.

Berman, Marshall. 1982. *All That Is Solid Melts Into Air: The Experience of Modernity*. New York: Penguin.

Bunker, Stephen G. 1985. *Underdeveloping the Amazon: Extraction, Unequal Exchange, and the Failure of the Modern State*. Urbana and Chicago, IL: University of Illinois Press.

Chatterjee, Partha. 2012. *The Black Hole of Empire: History of a Global Practice of Power*. Princeton: Princeton University Press.

Clark, Brett and John Bellamy Foster. 2009. "Ecological Imperialism and the Global Metabolic Rift: Unequal Exchange and the Guano/Nitrates Trade." *International Journal of Comparative Sociology* 50(3–4):311–334.

Cottrell, William Frederick. 1955. *Energy and Society: The Relation Between Energy, Social Change, and Economic Development*. New York: McGraw-Hill.

Cushman, Gregory. 2013. *Guano and the Opening of the Pacific World: A Global Ecological History*. Cambridge: Cambridge University Press.

Economy, Elizabeth. 2010. *The River Runs Black: The Environmental Challenges to China's Future*. Ithaca, NY: Cornell University Press.

Elvin, Mark. 1972. "The High-Level Equilibrium Trap: The Causes of the Decline of Invention in the Traditional Chinese Textile Industries." Pp. 137–192 in *Economic Organization in Chinese Society*, edited by W. E. Willmote. Stanford, CA: Stanford University Press.

———. 2004. *The Retreat of the Elephants: An Environmental History of China*. New Haven, CT: Yale University Press.

Frank, Andre Gunder. 1998. *ReOrient: Global Economy in the Asian Age*. Berkeley, CA: University of California Press.

Glahn, Richard von. 1996. *Fountain of Fortune: Money and Monetary Policy in China, 1000–1700*. Berkeley, CA: University of California Press.

Golas, Peter. 1999. *Chemistry and Chemical Technology, Part XIII: Mining* (Volume 5, Part 8). Cambridge: Cambridge University Press.

Hartwell, Robert. 1967. "A Cycle of Economic Change in Imperial China: Coal and Iron in Northeast China, 750–1350." *Journal of the Economic and Social History of the Orient* 10:102–159.

Harvey, David. 2005. *The New Imperialism*. Oxford: Oxford University Press.

He, Guangwei. 2014. "China's Dirty Pollution Secret: The Boom Poised Its Soil and Water." Available at http://e360.yale.edu/feature/chinas_dirty_pollution_secret_the_boom_poisoned_its_soil_and_crops/2782/.

Horowitz, Richard S. 2003. "Breaking the Bonds of Precedent: The 1905–1906 Government Reform Commission and the Remaking of the Qing Central State." *Modern Asian Studies* 34(4):775–797.

Jones, Christopher. 2010. "A Landscape of Energy Abundance: Anthracite Coal Canals and the Roots of American Energy Dependence, 1820–1860." *Environmental History* 15(3):449–484.

———. 2014. *Routes of Power: Energy and Modern America*. Cambridge, MA: Harvard University Press.

Klinger, Julie. 2015. *Mining Frontiers*. Unpublished Ph.D. Dissertation, Department of Geography, University of California, Berkeley.

Lee, En-han. 1977. *China's Quest for Railway Autonomy, 1904–1911: A Study of the Chinese Railway-Rights Recovery Movement*. Singapore: Singapore University Press.

Marks, Robert. 1998. *Tigers, Rice, Silk, and Silt: Environment and Economy in Late Imperial South China*. Cambridge: Cambridge University Press.

———. 2011. *China: Its Environment and History*. Lanham, MD: Rowan & Littlefield.

Moore, Jason W. 2015. *Capitalism in the Web of Life*. New York: Verso.

Nef, J. U. 1966. *The Rise of the British Coal Industry*. Hamden, CT: Anchor Books.

Parthasarathi, Prasannan. 2011. *Why Europe Grew Rich and Asia Did Not: Global Economic Divergence, 1600–1850*. Cambridge: Cambridge University Press.

Pomeranz, Kenneth. 2000. *The Great Divergence: China, Europe, and the Making of the Modern World-Economy*. Princeton, NJ: Princeton University Press.

———. 2002. "Political Economy and Ecology on the Eve of Industrialization: China, Europe, and the Global Conjuncture." *The American Historical Review* 107(2):425–447.

Richards, John F. 2003. *The Unending Frontier: An Environmental History of the Early Modern World*. Berkeley, CA: University of California Press.

Richthofen, Ferdinand von. 1903. *Baron Richthofen's Letters 1870–1872* (Second Edition). Shanghai: North China Herald Office.

Rowe, William T. 2001. *Saving the World: Chen Hongmou and Elite Consciousness in Eighteenth-Century China*. Stanford, CA: Stanford University Press.

Sachs, Jeffrey D. and Andrew M. Warner. 2001. "Natural Resources and Economic Development: The Curse of Natural Resources." *European Economic Review* 45:827–838.

Sieferle, Rolf Peter. 2001. *The Subterranean Forest: Energy Systems and the Industrial Revolution*. Cambridgeshire, UK: White Horse Press.

Schivelbusch, Wolfgang. 1986. *The Railway Journey: The Industrialization of Time and Space in the 19th Century*. Berkeley, CA: University of California Press.

Shapiro, Judith. 2001. *Mao's War Against Nature: Politics and the Environment in Revolutionary China*. Cambridge: Cambridge University Press.

———. 2012. *China's Environmental Challenges*. Cambridge, UK: Polity Press.

Shulman, Peter. 2015. *Coal and Empire: The Birth of Energy Security in Industrial America*. Baltimore, MD: Johns Hopkins Press.

Smil, Vaclav. 1994. *Energy in World History: Essays in World History*. Boulder, CO: Westview Press.

Sombart, Werner. 1902. *Der Moderne Kapitalismus* (Volume 2). Berlin: Duncker & Humblot.

Sugihara, Kaoru. 2003. "The East Asian Path of Economic Development: A Long-term Perspective." Pp. 78–123 in *The Resurgence of East Asia: 500, 150 and 50 Year Perspectives*, edited by Takeshi Hamashita, Mark Selden, and Giovanni Arrighi. London and New York: Routledge.

Tsing, Anna Lowenhaupt. 2005. *Friction: An Ethnography of Global Connections*. Princeton, NJ: Princeton University Press.

———. 2012. "On Nonscalability: The Living World is Not Amenable to Precision-nested Scales." *Common Knowledge* 18(3):505–524.

———. 2015. *The Mushroom at the End of the World: On the Possibility of Life in Capitalist Ruins*. Princeton, NJ: Princeton University Press.

Wallerstein, Immanuel. 1974–2011. *The Modern World System* (4 Volumes). New York: Academic Press.

Weber, Max. 1904/1958. *The Protestant Ethic and the Spirit of Capitalism.* New York: Charles Scribner's Sons.

Wong, Edward. 2014. "China Exports Pollution to US, Study Finds." *New York Times*, January 20, 2014. Available at http://www.nytimes.com/2014/01/21/world/asia/china-also-exports-pollution-to-western-us-study-finds.html?_r=0.

Wright, Tim. 2012. *The Political Economy of the Chinese Coal Industry: Black Gold and Blood-Stained Coal.* London: Routledge.

Wrigley, E. A. 1988. *Continuity, Chance and Change: The Character of the Industrial Revolution in England.* Cambridge: Cambridge University Press.

Wu, Shellen. 2012. "Mining the Way to Wealth and Power: The Late Qing Reforms of Mining Law, 1895–1911." *The International History Review* 32(3):581–599.

———. 2014. "The Search for Coal in the Age of Empires: Ferdinand von Richthofen's Odyssey in China, 1860–1920." *The American Historical Review* 119(2):339–362.

———. 2015. *Empires of Coal: Fueling China's Entry into the Modern World Order, 1860–1920.* Stanford, CA: Stanford University Press.

Zanasi, Margherita. 2013. "Western Utopias, Missionary Economics, and the Chinese Village." *Journal of World History* 24(2):359–387.

Part III

Thoughts on What Is Being Done? What Is to Be Done? And Who Should Do It?

10

Global Climate Justice Activism: "The New Protagonists" and Their Projects for a Just Transition

Jackie Smith and Jacqueline Patterson

The contributors to this volume have provided ample evidence for fundamental, transformative change in the world-system. If there remained any doubts, their analyses show that the capitalist world-system threatens not only the well-being of a majority of the world's people but also the very survival of our planet. The findings presented here validate claims that have been made by popular struggles for many decades, and we hope that their precision and robustness will help motivate more vigorous action for radical change.

In this chapter, we argue that the urgency of the ecological and economic conditions that many people now face and the immense inequalities that have grown and become more entrenched require that scholars

J. Smith (✉)
Department of Sociology, University of Pittsburgh, Pittsburgh, PA, USA
e-mail: jgsmith@pitt.edu

J. Patterson
National Association for the Advancement of Colored People Environmental and Climate Justice Program, Baltimore, MD, USA
e-mail: jpatterson@naacpnet.org

R. S. Frey et al. (eds.), *Ecologically Unequal Exchange*,
https://doi.org/10.1007/978-3-319-89740-0_10

move outside our familiar territory and embrace more deliberately the work of advancing social transformation and what activists are calling "just transition."[1] Using the knowledge we have about the outrageous injustices of the capitalist world-system, we need to be part of the project to imagine what a new world-system might look like and to identify ways to advance such a vision. Activist groups led by the most impacted front-line communities have long been working to do this, but their efforts must be supported by people from all walks of life.

Clearly "revolution" will not look like the version told in our history books. Success in taking over the state apparatus without larger transformation of the world-system is not likely to improve the economic or ecological prognosis for any given people in a significant way; in fact, most radical movements are not engaged in projects to take over the state. Addressing climate change requires dramatic actions at a global level to encourage and enable similarly dramatic changes in localities everywhere. For this to occur, more scholars must engage with the question, "What would a revolution look like today?" A necessary follow-up to that question is "*Who* will change the world, and *how*?"

We begin to offer some thoughts on these questions based on Smith's participatory research[2] and Patterson's active involvement with environmental justice (EJ) activists over more than a decade. We believe revolutionary change is emergent in movement spaces where people have long been developing shared analyses and theories of global social change and helping cultivate collective power and agency by building unity among a diverse array of activists, organizations, and movements. But the marginalization of radical movements and their discourses, and the related failure of scholars to engage fully in the task of helping prioritize the needs of the world's dispossessed and to legitimize and augment their voices has meant that this work remains invisible in both the academy and in mainstream thinking and discourse.

The preceding analyses leave us wondering about what pathways exist for altering long-term historical processes of ecologically unequal exchange and the global structures that reproduce these. What would a transformation of the existing world-system look like, and how might people take steps to move us in that direction? We look for answers to these questions to the groups that have worked most deliberately and urgently to advance system change, specifically the global EJ movement and its global networks. This movement has contributed to the emergence and spread of a number of concrete projects—and networks supporting these projects—that manifest practices that could fundamentally alter the global economic and political order.

The transformative projects discussed below reflect just some of the more prominent ones to emerge in recent years, and these projects continue to gain increased visibility and attention. They include food sovereignty, solidarity economies, and Human Rights Communities—projects that engage overlapping networks of environmental and economic justice advocates. These projects merit our attention as potentially system-transforming ones because they would—if widely adopted—undermine basic processes that are necessary for the capitalist world-system to function. In other words, they target those very processes that enable capital accumulation and thus the reproduction of the capitalist world-system. With these projects, the system's opponents not only prevent their own (further) dispossession by denying capital its ability to continue appropriating labor and resources from working people and communities, but they also help deepen the existing systemic crisis (see Chase-Dunn 2013). The latter is accomplished in part by exacerbating the crisis of profitability but more importantly by showing that there are indeed alternatives to capitalism, undermining the system's hegemony and legitimacy.

Marxist analysts have identified a number of processes deemed essential to capitalist accumulation and the sustained operation of the modern world-system. These include, among others, depeasantization, proletarianization, commodification, globalization, imperialism, and techno-managerialism. By empowering local agents, celebrating rural and indigenous cultures, resisting materialism and militarism, and privileging local knowledge and lived experience, the activist projects explored below serve to reverse basic logics and practices of globalized capital. They represent what Icaza and Vázquez refer to as "decolonizing, epistemic struggles" (Icaza and Vázquez 2013:689), or what Dalsheim views as forms of "counter-conduct" that help put forward multiple "heterotopias," or spaces where "hegemonic structures are *represented*, *contested* and *inverted*" (2016:36, emphasis in original).

According to Icaza and Vázquez (2013:684), "epistemic struggles," are where activists "are producing and theorizing other forms of the political, other economies, other knowledges" that lie outside dominant, anthropocentric market and state institutions. These struggles are seen as advancing more dignified life-worlds. Often such struggles remain less visible even in the scholarly research on social movements, since they go beyond the premises of modernity and operate in alternative epistemic and institutional frameworks (Conway 2016; Esteva and Prakash 1998). Thus, before we discuss the three types of transformative projects being advanced by global activists, we provide a brief and partial history of what has become known as the climate justice movement. This movement brings together the "new global protagonists" for climate justice, who have helped disrupt the stalemated climate negotiations and catalyzed the emergence of new counter-hegemonic alliances.[3]

The New Climate Justice Protagonists

The "new protagonists" of the climate justice movement are people from what are known as "frontline communities" engaging in direct action and other forms of protest in response to climate change and the forces driving it. Naomi Klein's 2014 book, *This Changes Everything*, has drawn popular attention to the diverse grassroots struggles of indigenous peoples and other communities most impacted by fossil fuel extraction and energy generation. And the 2014 Peoples Climate March in New York was a watershed in demonstrating the critical role and leadership of low-income communities of color in the climate justice movement. However, as we shall see, there is a long history of frontline community engagement to generate popular pressure and alternative projects that disrupt and help transform the existing, carbon-intensive system and allow for what activists call a "just transition" to a low-carbon society.[4]

The EJ movement has emerged in communities of color over recent decades in response to the ecological injustices explored in the contributions to this volume and to the long-standing and increasing inability of the capitalist system to provide for the basic needs and security of the people and communities who bear the greatest costs for the system's operation

(Lerner 2010). The global movement has emerged from several strands of organizing in different parts of the world, but in the United States it grew out of an explicit critique of the mainstream environmental movement, which had "blatantly omitted" the environmental claims being made by communities of color and indigenous groups (Taylor 2000, 2010:6).

Two important global developments in the early 1990s helped create the conditions that brought together and amplified the voices of people of color in global environmental politics, contributing to expanding the transnational conversations and connections that form the foundations of contemporary thinking and organizing around EJ and its powerful critique of global capitalism. First, by the 1990s, transnational organizing by indigenous groups had developed, in part through the repeated opportunities indigenous leaders had to meet, including formal United Nations (UN) meetings addressing discrimination against indigenous populations by member governments. These international networks helped convene the Indigenous Alliance of the Americas Continental Gathering, "500 Years of Indian Resistance," in Quito Ecuador in July of 1990. Representatives from 120 Indian Nations and organizations met for several days to develop a declaration that called for indigenous people to come together to defend their autonomy and control of their territories. They recognized the task as one that required a fundamental transformation of dominant institutions alongside deliberate efforts to build alliances with other groups who supported indigenous demands for autonomy:

> The achievement of this objective is a principal task for Indian Peoples. However, through our struggles we have learned that our problems are not different, in many respects, from those of other popular sectors. We are convinced that we must march alongside the peasants, the workers, the marginalized sectors, together with intellectuals committed to our cause in order to destroy the dominant system of oppression and construct a new society, pluralistic, democratic and humane, in which peace is guaranteed. (Indigenous Alliance of the Americas 1990)

The articulation of shared aims and alliance strategies brought together a globally networked community from a largely diverse array of indigenous groups whose primary allegiance and energies are firmly connected to their traditional lands. The 500th anniversary of Christopher Colombus' arrival in the Western hemisphere created an opportunity for indigenous peoples to come together as global actors to articulate their demands and find ways to resist further violence to their livelihoods and cultures. Global organizations and networks of indigenous peoples have, since this time, continued to expand and develop in coherence, strategic capacity, and influence (Becker 2011; Brysk 2000; Hall and Fenelon 2009; Morgan 2007; P. Smith 2016). Reflected in this document are some of the core ideas that have and that continue to guide today's EJ movements, including the need for leadership to emerge from those most impacted by capitalism; the recognition that the key struggle is over fundamentally different ideas about how to organize society and humans' relationship with the earth; the centrality of local autonomy and control over local territories; and the idea that changing the status quo requires work to build diverse alliances that support and amplify the leadership and voices from frontline communities.

A second development that helped focus and amplify the voices of people of color in global political arenas is the emergence in the years leading up to the UN Conference on Environment and Development of international discussions about the importance of biodiversity and about international strategies for its preservation. These conversations impinged directly on communities of color in particular, since the models for addressing biodiversity preservation being advanced by elites impacted directly the access of indigenous communities and other groups—often those in the global, people of color, majority—whose very livelihoods rely on access to land and natural resources. Such communities tend to utilize forests and other ecosystems in much more limited and therefore sustainable ways than do industrial societies, yet their access to traditional lands and commons was threatened by proposals for biodiversity preservation being discussed in global arenas (Alston and Brown 1993; Escobar 2008).

Simultaneously in the United States, practitioners working to analyze and address the racial disparities in public health risks in the United States were coming together with others working on racial and environmental justice to convene the National People of Color Environmental Leadership Summit in Washington, D.C. in 1991. This meeting resulted in the Principles of

Environmental Justice,[5] which articulated connections between racial inequities and environmental degradation and outlined steps toward EJ. Following the Summit, organizers attended and brought these Environmental Justice Principles to the UN Conference on Environment and Development in Rio, sharing their analysis and developing relationships with global activist networks (Chavis 1993). Significantly, the principles also provide guidance for multi-racial and cross-sectoral coalition-building. US activists that are part of EJ networks have remained exceptionally involved in helping connect grassroots organizing efforts in the United States with global sites, including both the World Social Forum process and the inter-governmental negotiations on the UN Framework Convention on Climate Change.

The fruits of some of these initial mobilizations around EJ can be seen most visibly in a number of later developments. First, in 2007 the inter-state system was shaken up by a major engagement of more radical EJ activist groups in the official negotiations around the UN Framework Convention on Climate Change. The involvement of these groups led to a split in the civil society alliances mobilizing around the global climate talks, and a vibrant new Global Climate Justice Alliance was born (Bond 2012; Hadden 2015). Second, following the failed climate talks in Copenhagen in 2009, Bolivian President Evo Morales hosted more than 30,000 representatives of governments and civil society groups at the World People's Conference on Climate Change and the Rights of Mother Earth in Cochabamba. Morales' move represented a dramatic shift from the status quo politics of inter-state relations.

By inviting popular movements to engage on an equal basis with government representatives, the World Conference challenged the prevailing power relations in inter-state politics and created a space for the development of a counter-hegemonic alliance of state actors and social movements (Building Bridges Collective 2010). Although the Conference was not well reported in the United States, its final, People's Declaration was a paradigm-changing statement that represented a radical shift in global political discourse. Not only did it name the global capitalist system as the cause of climate change, calling for its abolition, but it also made social movements, rather than states, the primary agents responsible for carrying forward the declaration. The document highlights many of the solutions to the climate crisis put forward by radical movements (WPCCC 2010).[6] Among the actions the Declaration calls for are popular referenda on how to respond to the climate crisis; attention to "climate debt" and reparations; the creation of a global climate tribunal to hold powerful countries and polluters accountable; advancing rights for Mother Earth; and replacing capitalism's growth logic with the indigenous notion of *buen vivir.* Although these demands are not likely to find much reception among the most powerful global actors—and indeed Morales' efforts to introduce these in the UN General Assembly have not, so far, advanced—the fact that they offer a real alternative to the capitalist logic creates room for popular "political imagination" and engagement with concrete ideas about how to address the climate crisis. Over recent years, movements have converged around some of these proposals and brought their transformative ideas from frontline communities to a much broader, global audience (Smith 2014).

Third, subsequent international climate negotiations have provided spaces where climate justice activists have come together to advance their thinking and strategic alliances, even as government talks have remained deadlocked over how to respond to the climate crisis (Bond 2012; Goodman and Salleh 2013:418). In spaces such as the World Social Forum, which hosted a 2012 Thematic Social Forum on Climate Change[7] in advance of the inter-governmental 20-year review conference called "Rio+20," activists came together to build a transnational identity and to unite their struggles around demands for environmental and climate justice. At the 2012 WSF Thematic Forum, activists countered the official "green economy" platform called "The Future We Want" with their own alternative, "Another Future is Possible"[8] (Goodman and Salleh 2013; Smith 2014). Subsequently, climate justice networks convened at the Climate Space at the 2013 World Social Forum[9] in Tunisia and again for the People's Climate March in New York in September 2014. In these spaces, activists have worked to refine and advance the kinds of projects discussed below that seek to respond to the needs of people and communities facing imminent threats from a changing climate. They have built a more coherent and cohesive shared analysis and sense of unity that has made them a presence in the global climate struggles. They have consistently put forward and helped translate for a culturally and class-diverse audience the ideas that have been integral to EJ from its very origins, including the idea that those most impacted by capitalism must lead the struggles for a more just world order (see, e.g., Patterson 2013; Podesta and Smith 2014).

Thus, the "new Climate Justice Protagonists" include actors previously invisible or marginalized in global debates, including peasants, indigenous peoples, immigrant workers, and urban communities (see, e.g., McKeon 2015; Salleh 2012). Together, they are advancing a theory and a strategy for changing the capitalist system. They are doing so in part by responding to the immediate needs and threats to livelihood they now face. But as they struggle for survival, they are experimenting with projects that illuminate paths to a more just and equitable as well as a more ecologically sustainable world-system for all. And they are working to build a broader movement for EJ and human rights.[10]

Contemporary Projects for World-System Transformation

As noted above, global capitalism depends upon its ability to exploit workers and the environment, and thus it has generated processes of depeasantization and urbanization that provide ready pools of workers for industry and to free up rural land for industrial uses. In addition, the system's energy-intensity and need for constant growth demands its continual expansion into ever more remote territories in search of new energy sources. This makes the system itself a perpetual threat to people and the environment—especially, and beginning with, those living in remote and often ecologically sensitive areas (Harvey 2009, 2012; Sassen 2014). Their long-term experiences of dispossession have made indigenous peoples a particularly powerful agent leading movements for global transformation. Not only do members of these "frontline communities" start from a position of having little to gain from the system and much to lose from its perpetuation, but as a result of the movement-building work that has been happening over several decades, they also bring a common and coherent set of alternative visions and practices that appeal to a wider population that is finally coming to recognize the inherent limits and contradictions of capitalism.[11] As these varied communities have come together to share their analyses and build their networks, they have found a source of unity and power as well as growing confidence from their complementary ideas about alternative and appropriate ways to organize human society (see Salleh 2012; Escobar 2015).

Inter-related processes of global capitalism—including globalization (or delocalization), proletarianization, depeasantization, commodification, and industrialization—not only are exceptionally threatening to indigenous people and people of color, but they also reinforce hierarchies and divisions among people and between humans and the earth. As analyses of unequal ecological exchange have made abundantly clear, such divisions facilitate the externalization of the social and environmental costs of capitalist production, displacing such costs away from those who benefit and onto the environment, communities of color, workers, and the larger society. Through various strategies, projects of the EJ movement help disrupt capitalism's competitive logic and the resulting ecologically unequal exchange by promoting cooperative practices that limit environmental impacts by, for instance, reducing the distances between sites of production and consumption, redefining development and core social values, and valorizing the work and identities that have been devalued by prevailing capitalist logics. These projects help advance community resilience by nurturing social cohesion and harmonious relations with the earth.

The following section provides brief summaries of some of the ways the EJ movement has responded to ecological and other threats. These initiatives are among the most prominent of those articulated by climate justice activists and related networks and organizations. Their prominence is reflected in the fact that they have attracted support from a large and diverse array of social sectors and activist networks, many of which first encountered these ideas through global activist spaces such as the World Social Forums. This observation alone demonstrates the critical importance to the work of global transformation of movements' *creation of autonomous spaces and networks* where counter-hegemonic and anti-systemic actors can converge and develop sustained mechanisms for communication and cooperation.

Food Sovereignty

The food sovereignty movement emerged in the 1990s through the leadership of La Via Campesina, a worldwide network of peasant organizations and small farmers that began in Latin America.[12] Food sovereignty advocates seek to transform the global food regime into a people-centered food system where all people enjoy "the right to sufficient, healthy and culturally appropriate food for all individuals, peoples and communities."[13] Food sovereignty activists demand the re-localizing control of land and food systems so that global food markets and global trade rules cannot undermine the ability of local producers and communities to shape decisions about their own subsistence and well-being.

Food sovereignty offers a profound challenge to the prevailing logics of global capitalism. At its core is an ecocentric rather than anthropocentric understanding of the world, which demands food systems that operate in harmony with natural systems. This logic contradicts capitalism's emphasis on industrialization. In addition, food sovereignty privileges human and non-material wealth over capital/material wealth. Essential to such a system is the valuing of food providers and the work of food production as well as a deep respect for the rights of farmers and other people and natural systems that contribute to the production of food. Ensuring the rights of food producers and consumers requires the localization of food systems and local democratic control over land and other resources, countering logics that have fueled urbanization. Local control over food systems requires expanding and valuing knowledge and skills related to food production and the cultures and social relations surrounding it. Thus, food sovereignty activists are both engaged in projects to advance models of localized food production while also advancing broader cultural and political movements against globalized capitalism.

Local knowledge and traditions can provide much-needed information about how communities can live in harmony with their respective ecosystems—it can help reverse the devastating impacts of the anthropocentric and consumerist logics of global capitalism. As expressed in the World Social Forum's (2012:19) *Another Future is Possible*:

> …food sovereignty, is designed as a comprehensive form of agricultural production that defends small-scale and indigenous farming to provide food, dignity, identity, and gender equality. These proposals also aim to nurture processes for the reconstitution of life territories and include demands for agrarian and fishing reforms that will once again give a key role to family farmers, fishing communities, their cultures, and ways of life. These proposals are articulated around three points: (1) family farmer and fishing knowledge, goods, and culture; (2) trading rights and regulations from the local to the global; and (3) joint participation and social oversight of the production system.

What is interesting in the ways activists articulate the notion of food sovereignty is how closely intertwined their understandings of food and food production are with culture and identity. In the above quote, we see that the ability to produce the quantity, quality, and types of food that are both nourishing and culturally appropriate is linked to basic human rights and dignity and to community.

Food sovereignty's emphasis on gender equality and ecocentrism defies the hierarchies of patriarchy and anthropocentrism that are integral to the capitalist world-system. In addition, it challenges the hegemonic logic that privileges the global over the local, urban over rural, and modern/industrial over traditional. Thus, although it embodies a set of concrete practices and strategies, food sovereignty has a significant cultural dimension that enhances its appeal to diverse constituencies and helps provide an ideological foundation that nurtures and reinforces counter-hegemonic practices and lifestyles.

The types of practices employed by food sovereignty advocates include, for instance, local seed banks; small-scale energy and irrigation systems; small-farmer cooperative and social organizations to support both production and distribution; urban buyers collectives and community supported agriculture initiatives; community gardens; research and extension efforts; among others (Figueroa 2015; Snipstal 2015).[14] Each of these practices, we argue, represents contributions to community resilience by placing greater control over food production and access directly in the hands of the people who are growing and consuming food. They enhance community food security while building and strengthening local markets and community infrastructures, and they counter global capitalism's logics of industrialized production and globalization by favoring more ecologically sustainable farming methods and reducing the distance between producers and consumers. By ensuring that consumers and producers share more direct community ties, this strategy reverses globalization's tendency to lengthen the distance between consumer and producer and thereby to externalize social and environmental costs. More localized production thus enhances working conditions and encourages environmental stewardship. It also makes producers and consumers less vulnerable to disruptions in global energy prices and supply chains, which are likely to increase in the face of energy scarcity, climate change, and related political instability.

In addition, food sovereignty helps valorize farming and small-scale production, countering capitalism's modernizing logic and discourses that stigmatize and devalue peasant lifeways. Indeed, a key element of food sovereignty strategy is the celebration of peasant farmers, who in the modernizing logic of capitalism were meant to become a relic of the past, their work being replaced by machines (McKeon 2015). Via Campesina's name translates as "peasant's way," demonstrating the conscious intention of food sovereignty activists to provide an alternative to global capitalism.

Thus, the concept of food sovereignty fundamentally challenges dominant ontologies and epistemologies by not just offering an alternative way of thinking about food systems but by reconceptualizing basic identities, cultural values, and social relations (see, e.g., Cormie 2016; P. Smith 2016). La Via Campesina challenges the dominant notions of peasants as artifacts of a pre-modern age and celebrates the traditions and cultures of actually existing peasants, promoting "repeasantization" as a solution to capitalism's multiple crises (McKeon 2015; Mann 2014). It also reinforces the values of living in harmony with the earth, local production, and traditional foods and practices—values which capitalist globalization rejects. This strategy is a direct response to the experiences of both rural and urban communities who have been dispossessed by processes of depeasantization (and its complement, proletarianization) and urbanization (McMichael 2008). The food sovereignty movement thereby valorizes the identities and the local knowledge of peasants and others who are part of what Goodman and Salleh (2013:421) refer to as the "meta-industrial class":

> Without doubt, the global majority of meta-industrial workers—urban women carers, rural subsistence dwellers, and indigenes—are hit hard by the exploitation and dispossession of ecological exhaustion. They also share the experience of exclusion and diminishment by social stratification and cultural bias. […] Yet, meta-industrials are victims only to hegemonic eyes. In a time of multiple crises, there is an urgent need for political decisions informed by ecologically embedded modes of existence. Women and men with 'holding skills' have a head start in constructing the parameters of a 'bio-civilisation' [….] As the focus of counter-hegemonic politics shifts from production to reproduction, 'another labour class' comes forward with unique capacities for regenerative knowledge.

In other words, the marginalization and exclusion of subaltern groups by the capitalist system has denied our society critical knowledge and experiences that are essential to our survival. Food sovereignty helps center the knowledge and voices of marginalized groups and to redefine values and priorities for a more just and ecologically resilient society. It redefines principles for producing and distributing food that reinforce community and environmental sustainability over markets and economic growth. Thus, food sovereignty is seen as a tool for social transformation and as a social process as much as a political platform (Snipstal 2015). Describing the Healthy Food Hub in an African American community on Chicago's south side, Figueroa concludes that food sovereignty projects are "not about 'chasing our piece of pie in the new green economy.' [They are], rather a point of entry into a larger project: to build forms of community wealth that can provide [marginalized groups] with much-needed autonomy and resilience against the forces that continue to lay waste to their communities" (Figueroa 2015:500).

The food sovereignty movement disrupts the logics and discourses that perpetuate global capitalism by centering human rights as a challenge to the prevailing order. Clayes (2015:456) calls food sovereignty a "full-fledged rights-based paradigm," which, according to McMichael (2015:437), "denaturalizes the 'global food system' by establishing (rights-based) claims of small producers to their own local food systems, which account for up to two-thirds of the world's food." Moreover, McMichael (2015:445) concludes, "combining a politics of rights and representation enables the construction of a counter-narrative to a monocultural development narrative, in a long-term crisis of unsustainability and inability to feed populations other than global consumers." It privileges local claims to land and its produce, challenging globalization's logic of scale at the same time as it valorizes the identities and lifeways of peasants and small-scale producers. As food sovereignty advocates generate practical alternatives to global capital, they are simultaneously building new cultural frameworks that both challenge the geo-culture of the capitalist world-system and help orient actors' decisions and actions around widely shared values. This contributes to their ability to mobilize diverse alliances and supportive constituencies while chipping away at the legitimacy of the existing order that subordinates human rights to material/economic goals.

Alternative Models for Economy and Society

The dramatic changes required to seriously reduce greenhouse gas emissions cannot be imposed in an authoritarian way, but rather they must be seen as necessary and legitimate. Thus, as Goodman (2009:511–512) observes, "climate change forces a wholesale re-democratisation of social relations, prefiguring new dimensions of economic democracy, intergenerational democracy, and transnational democracy." The projects we examine reflect this analysis and each of them advances more democratic political and economic practices and norms. Additional activist projects reflected in the work of frontline communities seeking to challenge globalized capitalism and advance more just and ecologically sound alternatives come under varying labels of solidarity economy, ecovillages, just transition, and human rights cities. As the terms applied to these projects implies, these initiatives help reorient the practices of participants and support *community and individual* survival through non-capitalist, democratic and egalitarian relationships and value systems. They challenge the competitive and discriminatory practices that are integral to global capitalism and present workable alternative models that are being enacted in communities around the world. As Escobar (2015:460) observes:

> The emphasis on the re-invention of communities is a powerful argument to deal with the amazingly pervasive practices keeping 'the individual' (anchored in markets and consumption) in place as the pillar of society and for imaging alternative regimes of relational personhood, in which personhood is also redefined within the *tejido* (weave) of life always being created with non-humans.

The projects described below, in addition to food sovereignty—which is often a key element of these other projects—reflect and articulate operating principles, values, and logics that support community-building and counter the logics of the prevailing capitalist order.

The notion of solidarity economy is probably the oldest of the examples provided above, and this project envisions and enacts economies based on cooperation, sharing, and on living with enough[15] rather than on competition, exploitation, and wealth accumulation. Solidarity economy projects include cooperatives, publicly owned banks, participatory budgeting, and other projects that facilitate production and exchange that reinforce community and ecological sustainability. They do so by re-embedding markets in communities and decommodifying exchange relationships. They thereby challenge capitalist logics of scale that fuel industrialization and urbanization and enable capital accumulation by separating people from their labor, land, and communities.

Ecovillage projects are intentional community models that prioritize social, economic, and ecological sustainability. Ecovillage participants seek to develop and institutionalize alternatives to ecologically destructive systems for the provision of transportation, food, energy, water, and waste-management. Inherent in this model is the belief that the breakdown of traditional forms of community, wasteful consumerist lifestyles, destruction of natural habitats, urban sprawl, industrial farming, and overreliance on fossil fuels are trends that must be changed in order to avert ecological disaster and create richer and more fulfilling ways of life. Ecovillages are small-scale communities that seek to minimize their ecological footprints and support alternative regenerative practices. Many advocates also seek independence from existing infrastructures, although others pursue more integration with existing infrastructure. Whether urban or rural, ecovillages tend to integrate community and ecological values within a principle-based approach to sustainability (Van Schyndel Kasper 2008). Johnathon Dawson (2006), former president of the Global Ecovillage Network, describes the five basic elements of ecovillages as: leading from the grassroots rather than governments; valuing and practicing community living; prioritizing community self-reliance for basic necessities such as food and water (vs. government/centralized support); nurturing a strong sense of shared values—often characterized in spiritual terms; and generating replicable models and educational experiences for others.

Explicit in the idea of ecovillages is that they can be replicated and scaled up. Indeed, as many participants quickly learned, achieving their goals requires changes in the larger set of relationships within a (bio) region. Thus, the vision of the EcoDistrict model is that of just, resilient and sustainable cities, from the neighborhood up. The concept of EcoDistricts is based on "urban regeneration and community development rooted in a relentless commitment to authentic collaboration and social, economic and ecological innovation that reimagines the future of cities" (EcoDistricts 2016).

The Just Transitions project is a more recent development, and it draws from these elements described above to bring groups together to support more concerted action to address the needs of communities that are being impacted by climate change. As its name implies, Just Transitions initiatives seek to ensure that the costs of climate change are not disproportionately borne by low-income people and people of color. As articulated in EJ networks, this project, perhaps more explicitly than the others described above, integrates an explicit rejection of the capitalist world-system and a conscious commitment to building an alternative system:

> Eliminating a socio-economic system requires a profound mass movement that changes socio-political systems and alters human behavior, particularly the behaviors that guide our collective choices about who decides what we produce and consume, what we produce and consume, why we produce and consume it, and why what we produce and consume is distributed in the unequal and inequitable manner that it is. In effect, we need a mass movement for a Just Transition and we have to build it![16]

In June 2013, the Climate Justice Alliance (CJA), a collaborative of more than 35 grassroots organizations in low-income and communities of color around the United States, launched the Our Power Campaign: Communities United for a Just Transition. The goal of the Our Power Campaign is to "bring together frontline communities to 'build the bigger we' for a just transition toward local, living economies."[17] The idea of just transition refers to the notion that the costs of shifting to a low-carbon society and the effects of climate change must be shared in a just and equitable way. CJA works to strengthen relationships between these frontline communities facing a variety of environmental threats and other sectors of progressive organizing, including environmentalists, labor unions, food sovereignty organizations, among others. Such alliances help raise public consciousness about the real costs of fossil fuel-intensive capitalist production on communities. Consistent with the EJ principles discussed above, the alliance works to ensure that people most impacted by economic and environmental crises lead efforts to resist and transform their conditions. The CJA organized assemblies at the US Social Forum in Detroit (2010) and sent delegations to international climate conferences, including those in the context of the UN and the World People's Conference on Climate Change and the Rights of Mother Earth, held in Bolivia in April, 2010. The Our Power Campaign grew from the discussions from these varied gatherings of activists and their engagements with other movements. A leading example of the application of Just Transition principles is the work led by Cooperation Jackson in Jackson, Mississippi. The Jackson Just Transition Plan incorporates the models of equitable and ecologically sustainable societies reflected in the ideas of solidarity economy, ecovillages, and human rights cities and outlines concrete goals and steps activists plan to take as they advance their vision of just transition.[18] As is unfortunately too frequently the case, residents of Jackson are motivated as much by the struggle for survival—a struggle that requires explicit attention to dismantling structural racism—as by value preferences for a system that is more just and that operates in harmony with nature.

The final example of projects for an alternative world-system is that of human rights cities. Human rights cities are "cities that explicitly refer to international human rights norms in their activities, statements or policy"

(Van den Berg and Oomen 2014:13). Such cities have been on the rise in recent years due partly to pressures caused by economic globalization such as migration and urbanization, financial crisis, and the devolution of state authority. Local authorities typically have the greatest influence over human rights protections. Yet, international human rights treaties are negotiated among national governments, and national authorities are ultimately responsible for their implementation. At the same time, globalization has put increased pressure on cities to compete for limited financial investments and to prioritize economic growth. In response to the new threats and opportunities at the local level, human rights advocates have been working to shift development discourse by demanding "rights to the city." The human rights city model offers mechanisms for holding municipal officials accountable to human rights standards that are widely resonant in the larger society. As communities face intensified pressures from the forces of globalization, such locally based movements advancing human rights claims are gaining momentum (Evans 2002).[19]

Recognizing that the prevailing capitalist system has done little to effectively address social problems such as poverty and social exclusion—and indeed that it creates and exacerbates these problems—human rights cities advocates contend that a human rights framework can help shift the debate away from competitive, market-oriented agendas that undermine social justice. It does so by mobilizing diverse community actors in support of a vision of a city that places social justice and community needs ahead of economic growth and "development." Human rights cities, like ecovillages, treat grassroots communities as the protagonists of change and agents of community survival and resilience. Of course, there remain important divisions among human rights advocates, and some models of human rights cities embrace reformist, individual rights-oriented approaches that do not threaten the prevailing capitalist order. However, the mobilizations of low-income people of color over recent decades have nurtured a vibrant and growing critical stream of human rights city organizing that is helping bring greater convergence to the human rights cities movement.[20] Building upon principles established and promoted by the EJ movement, human rights cities articulate demands for "people-centered human rights" that challenges conventional legalistic notions of rights and grounds rights claims in the needs of people and communities (Chueca 2016).[21] In practice, human rights cities engage residents in the collective work of envisioning a city based on the goal of maximizing human rights rather than profit. Long experience and documentation of environmental racism, moreover, has incorporated within human rights city organizing the idea that the protection of the natural environment (sometimes referred to "rights of Mother Earth") is integral to ensuring the full enjoyment of human rights.

Table 10.1 summarizes some of the main strategies or projects that are reflected in these strands of organizing we report on here, identifying the specific ways they help challenge the perpetuation of ecologically unequal exchange.

In sum, all of the projects we describe are examples of how social movements are modeling alternatives to capitalism and building "political cultures of opposition and creation" (Foran 2016). As Foran notes:

Table 10.1 Movement strategies and projects that disrupt environmentally unequal exchange

Project	Strategy	Implications for EUE
Food sovereignty	Enhancing local control of food production and distribution	Opposes capitalist appropriation of land and re-asserts "traditional" identities and cultures over modernist ones
Ecovillages	Enacting and promoting models of community living that reduce ecological footprints	Disseminates ecocentric ideology and inter-generational time frame; develops and supports alternative models and counter-hegemonic practices that maximize community and ecological well-being
Just transition	Building economic power of historically oppressed populations and connecting local movements with global climate justice networks	Reduces greenhouse gas emissions while challenging racial and class hierarchies. Fosters anti-racist, solidarity economy ideology and builds community capacity for collective action
Human rights cities	Organizing city policies around human rights principles/community well-being rather than markets/economic growth	Challenges hegemony of markets and economism in municipal policy and planning. Supports and disseminates alternative models of community governance

"Movements become even stronger when to a widely felt culture of opposition and resistance they add a positive vision of a better world, an alternative to strive for that might improve or replace what exists." As the social and ecological crises fuel opposition to the existing order, we may see expansion in the movements advancing these alternatives to capitalism and ecologically unequal exchange.

Discussion and Conclusion

Analysts of ecologically unequal exchange have provided ample evidence of how the modern world-system imposes disproportionate environmental costs and risks on less powerful groups—particularly those on the periphery of the world-system, people of color, and low-income people. In this chapter, we have documented how frontline communities experiencing the most harmful impacts of ecologically unequal exchange have long resisted systematic inequality and exclusion by developing projects to enhance EJ and community resilience. Such movements—often locally rooted—have contributed to the emergence of a global EJ movement that has wielded growing influence in recent years. Frontline communities have become the new protagonists of climate justice, articulating alternatives to capitalism that have attracted a growing array of adherents. By offering concrete ideas for reversing processes integral to the continuation of the capitalist world-system, and by privileging values and idea-systems that fundamentally challenge the geo-culture of the modern world-system, these actors offer promising insights into the question of "What is to be done?"

Nevertheless, however, appealing and compelling these models are, unless large numbers of people learn about them and have the means to participate, they will not alter the ecological or social crises we face. Moreover, efforts to promote cooperation and build social cohesion may become more complicated with the deepening of social and ecological crises. In addition, there remains the ever-present threat that movement projects will be coopted or subverted by elites, through schemes such as the "green economy." By appropriating movement language, elite forces can create the sense that they are addressing the crisis and produce both a reduced sense of urgency and confusion on the part of the general public.

Therefore, continued movement-building aimed at building diverse, multi-racial and multi-class relationships, and ongoing work to build the culture of opposition and creation is essential to enabling these projects to have the resources and support they need. Activist groups must continue to work at reaching "the middle" in order to bring transformative values and practices into the mainstream (Pastor and Prichard 2012). This requires creative attempts to develop communications capacity that can break through the mainstream corporate media monopoly to reach a wide range of people. Scholars can play essential roles working within movements to help activist networks build diverse coalitions that help create bridges among diverse groups and encourage mutual learning. They can also help movements develop strategic thinking and learning about how best to advance institutional and cultural change. Our experience working with movements reveals a need for greater support for the work of documentation, synthesis, and analysis of ideas and lessons generated from movement actions. And scholars' professional skills with communication can complement activists' own political communication skills to reach a broader public. Those hoping to reverse long-term processes of unequal ecological exchange can do so by helping draw more attention to the work of movements led by frontline communities and by contributing to efforts to better understand how their projects can be replicated and widely disseminated so that they nurture emerging alternatives to the capitalist world-system and a more just and ecologically sustainable world.

Notes

1. For a detailed discussion of what just transition work looks like, see "Climate Justice is Racial Justice is Gender Justice," Interview with Jacqueline Patterson *Yes! Magazine*, August 18, 2017 at http://www.yesmagazine.org/issues/just-transition/climate-justice-is-racial-justice-is-gender-justice-20170818.
2. Smith served on the National Planning Committee of the US Social Forum as a delegate from the International Network of Scholar Activists, as well as in local and national level efforts to help link global campaigns to more localized settings (see, e.g., Smith et al. 2011; Smith 2012).

3. Parts of this chapter draw from our contribution in *Resilience, Environmental Justice and the City*, Edited by Beth Schaefer Caniglia, Manuel Vallee, and Beatrice Frank, "Environmental Justice Initiatives for Community Resilience: Food Sovereignty, Just Transitions, and Human Rights Cities."
4. The notion of "just transition" first emerged from labor activists seeking to ensure that reducing the carbon-intensity of the economy did not disadvantage the most vulnerable workers. However, interpretations of just transition have varied between moderate and radical elements of the environmental justice movement. The groups of which we write embrace a more radical activist frame calling for large-scale social transformation that addresses both institutionalized racism and social exclusion while aggressively reducing greenhouse gas emissions (see Evans and Phelan 2016).
5. http://www.ejnet.org/ej/principles.html.
6. A second meeting called the World Peoples Conference on Climate Change and the Defense of Life was held in Tiquipaya in October 2015. An estimated 15,000 people attended that meeting, which was explicitly aimed to shape the Bolivian government's negotiating stance at the Paris climate talks later that year (see http://www.jallalla.bo/en/).
7. http://rio20.net/en/iniciativas/another-future-is-possible/.
8. http://rio20.net/en/iniciativas/another-future-is-possible/.
9. http://ggjalliance.org/road2paris.
10. For perspectives from leaders in this movement about the challenges of movement-building and cross-racial organizing, see Confronting Environmental Racism: Views from the Frontlines of the Climate Justice Struggle, January 22, 2015 at http://www.ucis.pitt.edu/global/climatechange_dialogueseries.
11. Of course, within these frontline communities there remain serious divisions over appropriate strategies, and often community leaders and members prefer efforts to benefit from participation in the prevailing capitalist order, including cooperation with extractive industries, over resistance.
12. http://viacampesina.org/en/.
13. http://www.nyeleni.org/spip.php?article290.
14. See also http://www.navdanya.org/.
15. There is resonance here with the Indigenous notion of *buen vivir* discussed above.
16. http://ggjalliance.org/just-transition-assemblies.
17. http://ggjalliance.org/ourpowercampaign.

18. http://www.cooperationjackson.org/blog/2015/11/10/the-jackson-just-transition-plan.
19. For more background on Human Rights Cities, see https://en.wikipedia.org/wiki/Human_Rights_City.
20. Smith has been part of an emerging network of human rights city leaders that has been convening within the framework of the US Human Rights Network. This network has recently become more formalized by creating a national steering committee and planning regular national Human Rights City convenings (see: http://www.ushrnetwork.org/our-work/project/national-human-rights-city-network).
21. http://www.ushrnetwork.org/resources-media/born-struggle-implemented-through-struggle.

References

Alston, Dana and Nicole Brown. 1993. "Global Threats to People of Color." Pp. 179–194 in *Confronting Environmental Racism: Voices from the Grassroots*, edited by Robert D. Bullard. Boston, MA: South End Press.

Becker, Mark. 2011. *Pachakutik: Indigenous Movements and Electoral Politics in Ecuador*. Lanham, MD: Rowman & Littlefield.

Bond, Patrick. 2012. *Politics of Climate Justice: Paralysis Above, Movement Below*. Durban, South Africa: University of Kwazulu-Natal Press.

Brysk, Alison. 2000. *From Tribal Village to Global Village: Indigenous Peoples Struggles in Latin America*. Stanford, CA: Stanford University Press.

Building Bridges Collective, editor. 2010. "Space for Movement: Reflections from Bolivia on Climate Justice, Social Movements, and the State." Leeds, UK: Footprint Workers Co-op. Available at https://spaceformovement.files.wordpress.com/2010/08/space_for_movement2.pdf.

Chase-Dunn, Christopher, editor. 2013. "Symposium: The Global System since 2008: Crisis of What?" *Journal of World-Systems Research* 19(2):175–201.

Chavis, Benjamin, Jr. 1993. "Forward." Pp. 3–5 in *Confronting Environmental Racism: Voices from the Grassroots*, edited by Robert D. Bullard. Boston, MA: South End Press.

Chueca, E. 2016. "Human Rights in the City and the Right to the City." Pp. 101–102 In B. Oomen, M. Davis, and M. Grigolo, editors. *Global Urban Justice: The Rise of Human Rights Cities*. Cambridge: Cambridge University Press.

Claeys, Priscilla. 2015. "Food Sovereignty and the Recognition of New Rights for Peasants at the UN: A Critical Overview of La Via Campesina's Rights Claims over the Last 20 Years." *Globalizations* 12(4):452–465.

Conway, Janet M. 2016. "The Study of Social Movements in the Modern-Colonial World System." Pp. 17–34 in *Social Movements and World-System Transformation*, edited by Jackie Smith, Michael Goodhart, Peter Manning, and John Markoff. London: Routledge.

Cormie, Lee. 2016. "Religious Others and Global Social Justice Movements." Pp. 132–158 in *Social Movements and World-System Transformation*, edited by Jackie Smith, Michael Goodhart, Peter Manning and John Markoff. London: Routledge.

Dalsheim, Joyce. 2016. "Other Moral Orders: Epistemology and Resistance—The Case of Israel/Palestine." Pp. 35–53 in *Social Movements and World-System Transformation*, edited by Jackie Smith, Michael Goodhart, Peter Manning, and John Markoff. London: Routledge.

Dawson, Johnathon. 2006. *Ecovillages: New Frontier for Sustainability*. London: Green Books.

EcoDistricts. 2016. *Vision, Mission, and Values*. Available at http://ecodistricts.org/about/vision-mission-values/. Accessed March 3, 2016.

Escobar, Arturo. 2008. *Territories of Difference: Place, Movements, Life, Redes*. Durham, NC: Duke University Press.

———. 2015. "Degrowth, Postdevelopment, and Transitions: A Preliminary Conversation." *Sustainability Science* 10:451–462.

Esteva, Gustavo and Madhu Suri Prakash. 1998. "Beyond Development, What?" *Development in Practice* 8(3):280–296.

Evans, Peter, editor. 2002. *Livable Cities? Urban Struggles for Livelihood and Sustainability*. Berkeley, CA: University of California Press.

Evans, Geoff and Liam Phelan. 2016. "Transition to a Post-Carbon Society: Linking Environmental Justice and Just Transition Discourses." *Energy Policy* 99:329–339.

Figueroa, Meleiza. 2015. "Food Sovereignty in Everyday Life: Toward a People-Centered Approach to Food Systems." *Globalizations* 12(4):498–512.

Foran, John. 2016. "Thoughts on Studying the Most Radical Social Movement of the Twenty-First Century, or after Paris: Building the Global Justice Movement of the Future." Paper presented at the annual meeting of the American Sociological Association, August 20, Seattle, WA.

Goodman, James. 2009. "From Global Justice to Climate Justice? Justice Ecologism in an Era of Global Warming." *New Political Science* 31(4):499–516.

Goodman, James and Ariel Salleh. 2013. "The 'Green Economy': Class Hegemony and Counter-Hegemony." *Globalizations* 10(3):411–424.
Hadden, Jennifer. 2015. *Networks in Contention: Global Civil Society and the Divisive Politics of Climate Change*. New York: Cambridge University Press.
Hall, Thomas D. and F. James Fenelon. 2009. *Indigenous Peoples and Globalization*. Boulder, CO: Paradigm Publishers.
Harvey, David. 2009. *Social Justice and the City*. Athens, GA: University of Georgia Press.
———. 2012. *Rebel Cities: From the Right to the City to the Urban Revolution*. New York: Verso.
Icaza, Rosalba and Rolando Vázquez. 2013. "Social Struggles as Epistemic Struggles." *Development and Change* 44(3):683–704.
Indigenous Alliance of the Americas. 1990. "500 Years of Resistance Declaration of Quito, Ecuador." Available at https://www.cumbrecontinentalindigena.org/quito90.php. Accessed September 24, 2015.
Klein, Naomi. 2014. *This Changes Everything: Capitalism Versus the Climate*. New York: Simon and Schuster.
Lerner, Steve. 2010. *Sacrifice Zones: The Front Lines of Toxic Chemical Exposure in the United States*. Cambridge, MA: MIT Press.
Mann, Alana. 2014. *Global Activism in Food Politics: Power Shift*. New York: Palgrave Macmillan.
McKeon, Nora. 2015. "La Via Campesina: The 'Peasants' Way' to Changing the System, Not the Climate." *Journal of World-Systems Research* 21(2):241–249.
McMichael, Philip. 2008. "Peasants Make Their Own History, but Not Just as They Please." Pp. 37–60 in *Transnational Agrarian Movements: Confronting Globalization*, edited by S. M. Borras Jr., M. Edelman, and C. Kay. Malden, MA: Wiley-Blackwell.
———. 2015. "The Land Question in the Food Sovereignty Project." *Globalizations* 12(4):434–451.
Morgan, Rhiannon. 2007. "On Political Institutions and Social Movement Dynamics: The Case of the United Nations and the Global Indigenous Movement." *International Political Science Review* 28(3):273–292.
Pastor, Manuel and Michele Prichard. 2012. "L.A. Rising: The 1992 Civil Unrest, the Arc of Social Justice Organizing, and the Lessons for Today's Movement Building." Vol. Los Angeles, California: Liberty Hill, USC Program for Environmental and Regional Equity. Available at https://dornsife.usc.edu/pere/larising/.
Patterson, Jacqueline. 2013. "And the People Shall Lead: Frontline Community Leadership in the Movement Towards a Sustainable Planet." NAACP Climate Justice Program. Available at http://www.usclimatenetwork.org/

resource-database/and-the-people-shall-lead-centralizing-frontline-community-leadership-in-the-movement-towards-a-sustainable-planet.
Podesta, P. J. and Laura Smith. 2014. "The Changing Face of Climate Change: Will the Leaders of the People's Climate March Now Lead the Movement?" *Slate.com.*
Salleh, Ariel. 2012. "Green Economy or Green Utopia? Recognizing the Reproductive Labor Class Post-Rio+20." *Journal of World-Systems Research* 18(2):141–145.
Sassen, Saskia. 2014. "Expelled: Humans in Capitalism's Deepening Crisis." *Journal of World-Systems Research* 19(2):198–201.
Smith, Jackie. 2012. "Connecting Social Movements and Political Moments: Bringing Movement Building Tools from Global Justice to Occupy Wall Street Activism." *Interface* 4(2):369–382.
———. 2014. "Counter-Hegemonic Networks and the Transformation of Global Climate Politics: Rethinking Movement-State Relations." *Global Discourse: An Interdisciplinary Journal of Current Affairs and Applied Contemporary Thought* 4(1):1–19.
Smith, Jackie, Scott Byrd, Ellen Reese, and Elizabeth Smythe, editors. 2011. *Handbook of World Social Forum Activism*. Boulder, CO: Paradigm Publishers.
Smith, Peter (Jay). 2016. "Spirituality and Global Capitalism—Contested Perspectives." Pp. 212–231 in *Social Movements and World-System Transformation*, edited by Jackie Smith, Michael Goodhart, Peter Manning, and J. Markoff. London: Routledge.
Snipstal, Blain. 2015. "Repeasantization, Agroecology, and the Tactics of Food Sovereignty." *Global Food Governance in an Era of Crisis-Special Issue: Mapping the Global Food Landscape* 2(2):164–173.
Taylor, Dorceta E. 2000. "The Rise of the Environmental Justice Paradigm: Injustice Framing and the Social Construction of Environmental Discourses." *American Behavioral Scientist* 43(4):508–580.
———. 2010. "Introduction." Pp. 1–27 in *Environment and Social Justice: An International Perspective*, edited by Dorceta E. Taylor. Bradford, UK: Emerald Group Publishing.
Van den Berg, E. and B. Oomen. 2014. "Towards a Decentralization of Human Rights: The Rise of Human Rights Cities." Pp. 11–16 in T. Van Lindert and D. Lettinga, editors. *The Future of Human Rights in an Urban World: Exploring Opportunities, Threats and Challenges*. Amsterdam: Amnesty International.
Van Schyndel Kasper, Debbie. 2008. "Redefining Community in the Ecovillage." *Human Ecology Review* 15:12–24.

World People's Conference on Climate Change and the Rights of Mother Earth (WPCCC). 2010. "People's Agreement." Available at https://pwccc.wordpress.com/support/. Accessed February 9, 2016.

World Social Forum. 2012. "Another Future Is Possible." Final document from the World Social Forum Thematic Forum on Capitalist Crisis, Social and Environmental Justice. Porto Alegre, Brazil: World Social Forum. Available at http://rio20.net/wp-content/uploads/2012/02/Another-Future-is-Possible_english_web.pdf. Accessed February 10, 2016.

11

Splintering South: Ecologically Unequal Exchange Theory in a Fragmented Global Climate

David Ciplet and J. Timmons Roberts

The Poor are not asking for charity [but] for the need for us to co-operate on an equitable basis. Now the rich claim a right to regulate the development of the poor countries. And yet any suggestion that the rich compensate the poor adequately is regarded as outrageous.

Malaysian Prime Minister, Dr. Mohathir Mohamad (1992:232, cited in Okereke 2006)

This work was supported by the Robert and Patricia Switzer Foundation and the Horowitz Foundation for Social Policy. This chapter was originally published under the same title in the *Journal of World-Systems Research* (www.jwsr.org) vol. 23, no. 1, 2017, pages 372–398, in a special issue of the journal entitled *Unequal Ecological Exchange* and edited by R. Scott Frey, Paul K. Gellert, and Harry F. Dahms. The authors retained copyright to publish the paper elsewhere under original agreement with the *Journal of World-Systems Research*.

D. Ciplet (✉)
Environmental Studies Program, University of Colorado, Boulder, CO, USA
e-mail: david.ciplet@colorado.edu

J. T. Roberts
Institute at Brown for Environment and Society, Brown University, Providence, RI, USA
e-mail: J_Timmons_Roberts@brown.edu

R. S. Frey et al. (eds.), *Ecologically Unequal Exchange*,
https://doi.org/10.1007/978-3-319-89740-0_11

Malaysian Prime Minister Dr. Mohathir Mohamad's speech at the 1992 Rio Earth Summit captures the main axis upon which the negotiations on climate change hinged for nearly two decades. Developing countries, who embraced an identity of the "global South," stood against proposals by those in the "global North" that did not recognize their rights to manage their own economies, the structural forms of inequality that inhibited their development, and their right to compensation for costs from dealing with climate change. This reflected a structural worldview, with roots in dependency and world-systems theories (Roberts and Parks 2006), which had in part been carried over and adapted from other historic platforms of developing countries, such as the New International Economic Order of the early 1970s (e.g. Krasner 1985; Rothstein 2015).

Scholarship on climate change has often characterized the politics regarding rights and responsibilities on greenhouse gas emissions as a struggle between states in the global North and global South (e.g. Gupta 1997; Roberts and Parks 2006). However, we argue here that given recent shifts in the contemporary world order and within the United Nations Framework Convention on Climate Change (UNFCCC), such analysis presents a static and no-longer accurate view of global environmental inequality. Scholarship also fails to clarify the alliances and conditions that structure possibilities for resistance. What is missing, we contend, is a nuanced understanding of the global South as a complex and changing set of relations reflecting shifts in the historic world order, and dynamics specific to the contemporary climate regime.

This weakness extends beyond works on climate change to the field of ecologically unequal exchange, which is the application of world-systems analysis to ecological relations between states and peoples. Scholarship on ecologically unequal exchange has largely focused on documenting the unequal distribution of environmental bads and goods in the world system (Hornborg 2001; Rice 2007; Jorgenson and Clark 2009; Shandra et al. 2009). Far less attention has focused on the historically specific political dynamics that shape the global governance of environmental inequality, including the implications of a politically and economically fragmented global South.

In this chapter, we ask, what do contemporary developments between global South states within the UNFCCC process reveal for theory about the governance of ecologically unequal exchange, and avenues for resistance?

We identify and discuss three areas of tension that have emerged within the G-77 coalition since the pivotal climate negotiations in Copenhagen in 2009, and which were solidified as part of the Paris Agreements in 2015. These include tensions within the global semi-periphery, tensions between the semi-periphery and periphery, and tensions within the periphery.[1] Through analysis of these tensions, we offer three important areas for improvements in ecologically unequal exchange theory. First, theory must better consider the role of the semi-periphery, and divisions within the semi-periphery, in reproducing ecologically unequal relations between societies. Second, theory should account for how fragmentation between the periphery and semi-periphery may produce distinct challenges for peripheral states to resist governance forms which intensify ecologically unequal exchange. Third, theory should better account for the ways in which ecologically unequal exchange as mobilized as a collective action frame aligns with or diverges from the real-world distribution of environmental goods and bads in the world system.

This analysis is informed by our 20 years of participant-observation research at the UN climate negotiations.[2] The article is organized in five steps. First, we discuss the relevant literature concerning the political dimensions of ecologically unequal exchange in global governance, particularly related to international climate change politics. Second, we outline how international climate politics were historically structured around particular ideas of inequality in the world system between the global North and South, how and why the old North–South alignments shifted in the pivotal negotiations in Copenhagen in 2009, and the major tensions in the global South relevant to the politics of ecologically unequal exchange during the post-Paris period. We conclude with discussion of what insights this analysis contributes to ecologically unequal exchange theory.

Ecologically Unequal Exchange as a Political Lens

Ecologically unequal exchange builds upon the Prebisch-Singer hypothesis which asserts that deteriorating terms of trade exist for countries that export raw materials (Prebisch 1950; Singer 1950). As a result, wealthy nations become richer by concentrating the benefits of these resources, while poor nations become further impoverished as their societies are transformed to deliver these resources to the developed nations at lowest price (Cardoso and Faletto 1979; Bunker 1985). As developed by the dependency or structuralist school, the global "periphery" was seen in a losing role in relation to the "center" or "metropole," where wealthy countries drew resources and cheap labor from around the world to manufacture high-value products they could export back to the periphery.

The terms *core* and *periphery* were adopted and elaborated by North American sociologists in the world-systems theory tradition (Wallerstein 2011; Chase-Dunn 1998). The terms speak to not only the international division of labor but also the ways in which surplus value from the transnational production of goods and services is concentrated unequally across geographies. As Arrighi and Drangel (1986:12) argue, "Core activities are those that command a large share of the total surplus produced within a commodity chain and peripheral activities are those that command little or no such surplus." World-system theorists then added a region to their conceptual apparatus that sat between the top and bottom countries: the semi-periphery, which represented the middle of the global division of labor, with both core and peripheral activities (Arrighi and Drangel 1986). The key characteristic of the semi-periphery was that it acted as a middleman between the core and the peripheral nations around it (Wallerstein 1979). These semi-periphery nations led the exploitation of the other countries in their regions to bring their resources to the world market, managing labor and investments there. In doing so, they developed decidedly bimodal or mixed economies, with extremely modern sectors and vast internal regions continuing to live in pre-modern conditions (Hecht, Anderson, and May 1988).

World-system theorists have long argued that while the structure of the world system has been relatively consistent over time, individual states can and do move up or down in the hierarchy (Arrighi and Drangel 1986:28). Arrighi and Drangel (1986) argue that semi-peripheral states seek to exploit distinct advantages of their position for gains in the world system. Specifically, they "resist peripheralization by exploiting their revenue advantage vis-à-vis peripheral states and their cost advantage vis-à-vis core states" (Arrighi and Drangel 1986:27). They point to domestic strategies related to managing their position in global commodity chains as the primary mechanisms through which they attempt to do this.

Notably absent from this discussion are strategies pursued by semi-peripheral states to maintain or enhance world-system position through multilateral governance processes. That is, scholarly attention should be directed to how semi-peripheral states actively seek to change the rules of the global system through political action in multilateral fora.

Ecologically unequal exchange has built from understandings of structurally conditioned unequal exchange in commodities, pricing and labor, to unequal access by wealthy countries to natural resources, ecological well-being, and sink capacities in poor countries (Frey 2015; Hornborg 2001; Rice 2007; Jorgenson and Clark 2009; Shandra et al. 2009). For example, Rice (2007:43) defines ecologically unequal exchange as "the increasingly disproportionate utilization of ecological systems and externalization of negative environmental costs by core industrialized countries and, consequentially, declining utilization opportunities and imposition of exogenous environmental burdens within the periphery." Counter to ideas of ecological modernization that posit a delinking of capitalist growth from environmental degradation in "modernized" societies (e.g. Mol and Spaargaren 2000), ecologically unequal exchange scholars argue that ecological harm is externalized by wealthy countries onto poor ones, and ecological well-being is expropriated from them. Importantly, it is argued that these processes of inequality related to the environmental issues such as agriculture, mining, and energy are sustained by global systems of governance and elite-controlled networks, institutions, and organizations (Downey 2015).

Others have made the case that climate change is a case of ecologically unequal exchange, with peripheral countries not benefiting from the fossil-fuel intensive development of the core, while experiencing the ecological impacts first and worst. From this viewpoint, the disproportionate impacts and vulnerability in the periphery to climate change are understood as not merely a geographical anomaly, but as conditioned by a colonial history of unequal insertion into the world-economy and uneven trade relations (Roberts and Parks 2006) as well as "double exposure" to climate vulnerability and the detriments of economic globalization (O'Brien and Leichenko 2000).

In terms of the role that the semi-periphery plays in ecologically unequal exchange theory, Burns, Kick, and Davis (2003:362) argue that a pattern may exist of "recursive exploitation," whereby a nation in the "semiperiphery is at a disadvantage to one in the core, yet is able to work exchanges in its favor when they involve the semiperiphery or periphery." In other words, due to their position within the world order, semi-peripheral states may be in a position to mediate some of their environmental burden by dumping it upon states with a less favorable position (Rice 2007). However, this capacity has not always been supported empirically; in the case of deforestation, semi-peripheral countries have experienced higher rates of deforestation than those at the periphery, likely attributed to a historical artifact, whereby peripheral countries were yet to experience similar levels of urbanization (Burns, Davis, and Kick 1997; Burns et al. 2003; Jorgenson 2004). Studies have found that such relationships may vary by pollutant; for example, greenhouse gas emissions are linearly related to position in the world-system hierarchy, whereas methane emissions tend to be heaviest in semi-peripheral countries (Burns et al. 1997; Jorgenson 2004). In terms of the impacts of climate changing emissions, the states that are most peripheral in the world system, such as the 49 least developed countries (LDCs), are far more vulnerable to climate-related disasters than the global average, despite contributing almost nothing to its cause (Ciplet et al. 2013a). We have not seen explicit comparisons of climate change vulnerability between peripheral and semi-peripheral states, but most categorizations place the poorest nations in the lists of the most vulnerable (see e.g. Roberts and Parks 2006). Still, limited attention has been devoted to uncovering descriptive or causal relationships of ecologically unequal exchange between the semi-periphery and periphery, or even the core and semi-periphery.

Moreover, despite a growing literature discussing the empirical and theoretical dimensions of ecologically unequal exchange, there has been minimal attention to the contentious real-world political aspects of ecologically unequal exchange in practice, particularly as they take shape within changing global governance regimes. One recent exception is the work of Downey (2015), who analyzes the role of elite-controlled transnational networks in structuring global environmental inequality through governance institutions such as the World Bank, World Trade Organization, International Monetary Fund, and corporate-controlled commodity chain networks. However, he gives limited consideration to divisions in the non-core, and to agency of elites outside the core.

As for resistance to global environmental inequality, numerous studies highlight forms of resistance by peripheral states and civil society actors to the disadvantageous rules in global governance regimes on issues such as forests (Schroeder 2010; Ciplet 2014), biodiversity (Escobar 1998; Shiva 1996), waste (Okereke 2006), and climate change (Pettit 2004; Terry 2009; Roberts and Parks 2009; Ciplet 2014, 2015; Ciplet, Roberts, and Khan 2015). However, a few studies have explicitly linked these politics of resistance and the forms that they take in particular historical periods to conceptions of ecologically unequal exchange. What focus does exist in the literature has identified ecologically unequal exchange politics as hinging largely on an axis between North and South. Notably, Roberts and Parks (2006) argued at length that the roots of the G77 coalition unity in the UNFCCC politics lay across many issues far beyond the climate talks: in these nations' lack of access to meaningful participation in the global order, the deep inequity in their well-being compared to the wealthy nations, and their agenda for Third World solidarity. Several scholars have also identified the emergence of the concepts of "ecological debt" and "climate debt," informed by world-systems analysis and conceptions of ecologically unequal exchange, as frames of resistance adopted by peripheral states and civil society groups in the negotiations since 2000 (Bond 2010, 2012; Klein 2010; Roberts and Parks 2007, 2009; Ciplet 2015; Pickering and Barry 2012; Botzen, Gowdy, and van den Bergh 2008; Chatterton, Featherstone, and Routledge 2013). This perspective argues that the global North should remunerate the global South for a debt as the result of disproportionate polluting of the global atmosphere and its unequal consequences.

We have also pointed to changing dynamics within the world system relevant to the UNFCCC process, most notably, the hegemonic competition between the United States and China and its implications for international cooperation on climate change (Roberts 2011; Ciplet et al. 2015). While discussing the increasingly prominent role of emerging powers such as China, India, Brazil, and South Africa in the negotiations in Copenhagen and Cancun, Hurrell and Sengupta (2012:463) caution that it is important not to "underplay the continued relevance of understanding climate change within the North-South frame." We agree. This frame is still a major axis in the negotiations. However, it is far from the only one now of relevance. While scholars have brought attention to the shifting power between wealthy states and rising economic and "emissions powers" in the semi-periphery, much less attention has been directed to what this and other developments mean to unity in the global South and to the reproduction of inequality in global environmental governance.

The Old World Climate Order

The Group of 77 and China (G77) is a bloc of developing nations now numbering over 134 countries. As Vihma (2010:4) put it, the G77 is "a product of the North/South divide and the political economy of the late Twentieth Century. It is broadly based on a 'self-definition of exclusion' from world affairs." That is, the vast global South, consisting of all of Latin America, Africa, and nearly all of Asia, felt that they had been left behind over decades of efforts at economic development and globalization (Najam 2005; Roberts and Parks 2006). Brought into the world-economy through colonial conquest and continuing to be dependent on the production and export of minerals and agricultural products whose prices fluctuated wildly or tended to go downward, they saw themselves as trapped in structurally disadvantaged positions. These are the underlying forces that held the coalition together until Copenhagen in 2009, despite their diverging material interests.

At the beginning of the climate negotiations in the early 1990s, the G-77 was a largely reactive coalition because of its suspicion of the environmental negotiations as an agenda of the industrial countries. Poorer nations expressed that green concerns were a ruse to keep them poor, a conscious or unconscious effort by the wealthy nations to keep the poor nations from usurping their place atop the global hierarchy (Roberts and Parks 2007; Gupta 1997). The G-77 shared interests in pressuring the historically wealthy or developed countries (what are called "Annex 1" countries in the negotiations) to act according to their historical responsibility for having created the problem and their capabilities to address it (their wealth). Developing countries also advocated to maintain their own sovereignty from outside intervention (especially from limits on their ability to pursue national economic development), and for the provision by wealthy countries of adequate funds and the most modern technologies to help them deal with climate change.

Addressing climate change means reducing consumption of cheap fossil fuels and switching to what have historically been more expensive sources of renewable energy like wind and solar; it also can mean not clearing rain forests to create farmland to expand the national economy, and so on. For this reason, the G-77's initial approach to this new agenda was wait and see, learn and react, or resist and reject (Najam 2005). If they were to address climate change and other environmental concerns, they needed to be compensated for lost economic gains and helped with new green technologies. When it came time to draft the UNFCCC before the 1992 Rio Earth Summit and later as part of the 1997 the Kyoto Protocol, the G-77 and China succeeded in their goal of avoiding responsibility for making emission reductions.

To be clear, the G-77, which incorporates the periphery and semi-periphery nations, has never been a homogenous bloc, or one without conflict (see Vihma, Mulugetta, and Karlsson-Vinkhuyzen 2011). A key tension in the group from the start has been between the Alliance of Small Island States (AOSIS) and the Organization of the Petroleum Exporting Countries (OPEC). At the first meeting of the Conference of Parties (COP) in Berlin in 1995, when a majority of G-77 countries supported binding reductions of emissions, OPEC advocated against them, even for the industrial countries (fearing they might be next). The G-77 took stands against any taxes on carbon, insisting instead that they should be compensated for lost business since measures to respond to climate change would severely affect their economies by slashing their ability to sell oil. The idea of compensation of oil producers for lost revenue is enshrined in Article 4.8 of the Convention, which included special consideration for economic vulnerability to climate change response measures.

As for AOSIS, since the beginning of the climate negotiation process in 1989, this negotiating bloc was very active in attempting to insert binding commitments for greenhouse gas emissions reduction under the newly established Intergovernmental Negotiating Committee. AOSIS was particularly active in demanding ambitious, science-driven, legally binding emissions reductions targets and compensation funding for climate impacts. The group was the first to propose a draft text during the Kyoto Protocol negotiations calling for cuts in carbon dioxide emissions of 20 percent from 1990 levels by 2005 (Earth Negotiations Bulletin 1995). The group demanded the establishment of an international insurance pool for climate victims; it took ten years just to get loss and damage on the agenda in Cancun in 2010 and another three years just to begin a work program to research the issue.

The compulsion of AOSIS was obvious. The group's 44 members are spread across the South Pacific, Indian Ocean, and the Caribbean, Africa, the Mediterranean, and the South China Sea. AOSIS's unity comes from the fact that more than nearly any other countries, their physical survival as states is at stake due to steadily accelerating sea-level rise from climate change. The first report of the Intergovernmental Panel on Climate Change (IPCC), published in 1990, indicated an ominous development: sea-level rise due to climate change would condemn many low-lying areas to be submerged. In this effort, AOSIS found a willing partner in the European Union (EU), which, being influenced by public opinion and strong social movements, also showed great interest in controlling greenhouse gases from the beginning. Yet small island developing countries continued to stand behind G-77 statements and positions in the negotiations, which were generally for slowing the progress of aggressive climate treaties. Even those nations with quibbles about this position did so because their voice was so easily ignored when they spoke alone: if they could get some of their positions into G-77 statements, they had some chance of influencing a treaty.

Later in Bali in 2007, with the end of the first commitment period of the Kyoto Protocol in sight, the G-77 stood strong in negotiating a successor treaty that maintained a structurally divided view of the world. Most central, the Bali text cemented different expectations for the developed and developing countries—"a Bali firewall" that would be defended for years by many developing countries (Smith 2010; Ciplet et al. 2015). Nowhere did the Bali action plan describe whether or how countries might move from one group to another, either up or down. Nor was there clarity on how a scientifically adequate solution might be met, or clear rules for compensation for countries losing revenue from reducing their emissions sharply.

The New World Climate Order in Copenhagen

It wasn't until the pivotal negotiations in Copenhagen in 2009 that the G-77 would dramatically splinter. Perhaps most devastating to the unity of the G-77 was the formation of the coalition of Brazil, South Africa, India, and China, known as BASIC, in October 2009, just before the Copenhagen conference. At the time of their collaboration, these countries were highly diverse in their interests. Their economic base, energy infrastructure, and emission levels all varied greatly, as did the nature of their states and their approaches to making and meeting greenhouse gas emissions reduction goals. Nevertheless, the key moment at Copenhagen was when President Barack Obama of the United States joined with leaders of the BASIC coalition to draft their own climate deal, which completely set aside the existing negotiating texts. The draft mentioned the goal of keeping global mean temperatures under 2 degree Celsius rise but avoided any binding emissions reduction targets to achieve that and any mention of the time when perilously rising emissions must peak (Ciplet et al. 2015).

Most crucially, the Copenhagen Accord that they drafted entirely shifted the approach taken by the global community in the face of climate change. The earlier Kyoto Protocol approach was top down, with binding national commitments based on levels of emissions and capabilities of countries (usually understood to be roughly their level of income per capita). The Copenhagen approach that the United States and BASIC put forward was entirely voluntary and bottom-up, with nations pledging and reviewing their own choice on what emissions reductions they would undertake.

China and the United States, a rising and a declining hegemon that together emitted about 40 percent of all greenhouse gases on Earth, consciously avoided a time frame for a midterm emissions reduction target

(Roberts 2011). The bold move at Copenhagen showed the ascendant power of the BASIC group and its ability to work directly with the United States and to cut their G-77 colleagues and the EU out of the decision making. The way the Copenhagen Accord was cobbled together was unprecedented, for heads of state and governments rarely get directly engaged in, let alone lead, international climate change negotiations. The accord was quickly brought to a hand-picked group of 28 countries to rubber-stamp, with almost no time to review it thoughtfully and no opportunity to revise it (Ciplet et al. 2015). In this group of 28 were nearly all the wealthy OECD countries and just one representative from each of the developing world regions: Africa, Latin America, AOSIS, and Asia.

The new voluntary nationally determined approach in the Copenhagen Accord faced strong resistance from numerous leaders of peripheral states on both procedural and content grounds. The final all-night plenaries at Copenhagen were fiery, with a few feisty speeches by the countries willing to risk upsetting the global order and the ire of major aid and investment players, the United States and China. This accord and work by Mexico to formalize them in the 2010 Cancun Agreements paved the way for a bottom-up, voluntary approach to international mitigation that was adopted as part of the Lima Agreements in 2014 and the Paris Agreements in 2015, in which countries all brought their "Intended Nationally Determined Contributions" (pledges), or INDCs. During this period, the G-77 coalition further fractured along several lines. A series of new coalitions also emerged within the global periphery, some with competing identities and interests.

Splintering South: Three Fissures in the G-77

Tensions Between the Periphery and Semi-periphery

Since Copenhagen, the terms of what constitutes the "global South" have been under contention. One emergent tension within the G-77 has been between state coalitions such as the LDCs and the AOSIS on the one hand and rising industrial powers in BASIC on the other, especially about who should be required to commit to emissions reductions within the new NDC framework solidified in Paris. The moral force of a peripheral nation's extreme vulnerability to climate change is now often pitted against the need for development in emerging economies. For example, at one of the key informal meetings in the 2011 Durban negotiations, in response to the Indian environment minister's statement arguing for their right to development for meeting basic needs, the delegate Karl Hood from Grenada, representing AOSIS, reportedly retorted, "While they develop, we die; and why should we accept this?" (Roberts 2011).

Indeed, it is no longer possible to solely or primarily blame the global North for rising emissions. Developing countries now outpace developed countries in current carbon emissions (Center for Global Development 2015). The clear majority of projected emissions growth in the next two decades is expected in developing, not developed countries (Energy Information Administration 2013). In 2007, China surpassed the United States as the largest current global polluter, but remained far behind in terms of its cumulative historical emissions (Vidal and Adam 2007). But this is changing too. China surpassed the United States around 2015 or 2016 in terms of cumulative emissions (Doyle 2015) and has already overtaken the EU in emissions per capita (McGrath 2014).

In response to this new reality, there has also been a notable shift in messaging among the poorest countries, which are now beginning to call for a more sophisticated and historically relevant differentiation of responsibility between states, including those in the South. In 2015, prior to the Paris Negotiations, the LDCs negotiating group's official submission written by Nepal argued that the new framework should take "full account of current socio-economic realities" and be a single regime "applicable to all" (Nepal 2015:1). They argued that over the past 20 years the economic conditions in the world had considerably evolved, leading to changes between countries, including the current annexes of the Convention (Nepal 2015:4). Specifically, they called for "allowing some differentiation for developed countries, emerging economies, middle-income countries, the most vulnerable and the least developed countries based on agreed criteria" (Nepal 2015:1). While this may sound like a common-sense proposal, it is a stark departure from supporting proposals that maintain a rigid "North-South" divide that was enshrined in the Bali Firewall in 2007.

A new negotiating group called the Independent Association of Latin American and Caribbean Countries (AILAC), which officially launched itself in 2012 at the Doha negotiations, has also taken positions that challenge earlier G-77 convention (Ciplet et al. 2015). AILAC is notable in that it largely embraced the new voluntary "pledge and review" approach to emissions reductions, while several coalitions such as the LDCs and the AOSIS were still demanding a second commitment period of the Kyoto Protocol. The coalition, made up of Colombia, Costa Rica, Chile, Peru, Guatemala, and Panama, with the support of the Dominican Republic, viewed itself as being a bridge between conflicting North–South interests in the negotiations, and sought to encourage a more ambitious agreement by committing to action themselves. Specifically, the AILAC countries decided to stop waiting for emissions reductions or financial support from wealthy countries like the United States, and launch an ambitious case for low-carbon development at home and abroad. This decision was a major break from G-77 solidarity.

AILAC has downplayed the class-based identity of global South which is embraced by many peripheral states in the negotiations as structurally disadvantaged and deserving of compensation. They have instead focused on how developing countries can take responsibility themselves. For example, Peru made the first formal pledge for emission reductions by a Latin American country in 2015. As former Costa Rican advisor Monica Araya told El País, "[The negotiations are] always told as a battle of North versus South … but each time this explains less and less of what's happening" (Méndez 2012). She continued, "There is an alliance of countries that want all nations to take on binding obligations, and that the negotiations process is adapting to a changing world" (Friedman 2013a). Isabel Cavelier, a former negotiator for Colombia said in 2012, "We think we can show the world that we are developing countries, we have a lot of problems at home, but we are ready to act. If we can show that we can take the lead, and we're not waiting for the rest of the world, then we can [set] an example" (Friedman 2013b). AILAC negotiators are quick to point out that its positions do not undermine the core positions of the G-77 on equity, but they emphasize a more flexible interpretation of countries having to act according to their historical responsibility for climate change (CBDR+RC), to encourage all countries to commit to reducing their emissions.

A further fracture within the G-77 between the peripheral and semi-peripheral states occurred when, three days before the end of the Paris negotiations, a "high-ambition coalition" emerged in Paris, comprised of 79 African, Caribbean, AILAC, and Pacific countries, along with the EU and later the United States and Brazil, but without the other major global South powers, including China and India. The coalition formed in secrecy prior to the negotiations, and called for a legally binding agreement, a long-term goal on global warming commensurate with science, a review mechanism to assess emissions commitments, a unified system for tracking countries progress on meeting their goals, and eventually, a more ambitious emissions target of 1.5 degree Celsius temperature change (Mathiesen and Harvey 2015). Several of these demands contradicted the expressed interests of China and India. A previous "high-ambition" coalition that crossed the North–South divide had come together in Durban in 2011 between members of the LDCs, AOSIS, and the EU, at that point, to achieve a second commitment period of the Kyoto Protocol.

Tensions Within the Semi-periphery

In addition to rifts between the periphery and semi-periphery within the G-77 coalition, there are also many relevant tensions within the semi-periphery itself. The rising industrialized states are a very diverse group in terms of their emissions, economic activities, regional relations, and energy possibilities. For example, Turkey has the second-highest energy consumption growth after China and is dependent upon Russian and Iranian oil and gas and has plans to double its coal power capacity in the next four years (Friedman 2015). Brazil, for its part, depends far more on hydroelectricity and biofuels to power its growth. This makes Brazil far more efficient than many other states in its economic class in terms of carbon emissions per unit of Gross National Product (GNP). From a climate perspective, Brazil's main concern is to definitively control deforestation, which has been its largest source of its carbon dioxide emissions since the late 1980s. However, its commitment to lower its carbon emissions appears to be weakening (Edwards and Roberts 2015). Clearly, one should not assume that these two countries, or other rising or middle-income economies, will share the same positions in the post-Paris period, in which all are responsible for taking mitigation action, but have discretion on what forms their own commitments should take. Their distinct characteristics and interests uniquely shape their positions in regard to structures of ecologically unequal exchange in the world system.

While much could be written about fragmented interests between numerous countries that occupy this middle position in the global class structure and division of labor, here we focus on two of the major players, China and India. While they are often grouped together, China and India are in very different situations, and have taken decidedly different approaches in the contemporary negotiations, including in the Paris talks.

China has been more willing than India to take mitigation action commensurate with the demands of AOSIS and the LDCs, including providing financial assistance. China in the 1990s and early 2000s was very different from China today. Economically, its 7–10 percent annual growth and state-led capitalist transition has rocketed the nation to the highest levels of economic power. By some measures, China has just surpassed the United States and is now the world's largest economy, it is the workshop of the world in manufacturing, and it already is the holder of the world's greatest currency reserves and of other nations' debt (Katz 2014; Schiavenza 2014). China has increasingly seen climate negotiations as an important area of foreign policy to show that it is capable of addressing global problems and as an avenue for asserting leadership among developing countries (Chayes and Kim 1998). For this reason, China from the beginning worked for a united "G-77 and China" strategy (Economy 1997), perceiving its own role as speaker for the group (Heggelund 2007).

China is heavily investing in renewable and nonrenewable energy resources and infrastructure development in Asia, Africa, and Latin America. Unlike countries in the West, it is reported to have declined to make its investments conditional based on government reform, which makes it popular among a wide group of states (Alessi 2012). China is also a contributor of climate-related finance to many developing countries, particularly in Africa and Latin America. It seems likely that China's involvement as an investor and donor is responsible for some of the recipient countries' supportive responses to Chinese positions and leadership in climate change negotiations (Edwards and Roberts 2015).

Importantly, in November 2014, China agreed to a joint announcement with the United States to mitigate climate change, representing an important political breakthrough and for China a move beyond simplified notions of a North–South divide in responsibility. It also showed China's self-identification alongside a superpower, not making joint announcements with its BASIC or other G-77 partners. China also made major commitments to development assistance and investment (Hart, Ogden, and Dotson 2015; Khor 2015).

Thus, China went into the pivotal Paris negotiations, attempting to assume a position of global leadership, rather than that of merely an antagonist to the process. The Paris negotiations could have gone off the rails in the final minutes as the United States objected to key wording in the final document; China was reported to have stuck by the United States and not the G-77. But rather than fully distancing itself from its supposed peer group in the South, in Paris, China's President Xi Jinping continued to attempt to align with the interests of the weaker countries in the G-77, to call on wealthy states to scale up their climate finance and provide stronger support to developing countries (Mauldin 2015).

However, China's mitigation actions may fall far short of being adequate, due to its enormous net footprint and relatively high per capita emissions, and its hesitance in abandoning its fossil-fuel infrastructure. In addition, China's climate assistance may take the shape more of colonialism, as it gathers up land and resources in Africa and Latin America for biofuels, and uses its own companies to construct massive infrastructure projects. China's investments do not likely match a vision of compensation for damages based on ideas commensurate with "climate justice." Overall, while many commended China's action in the recent negotiations, others have been critical of the country's scale of ambition, the mechanisms of transparency in the country to achieve its stated goals, the scale of recent investments in fossil-fuel infrastructure, and its underreporting of previous emissions.

As for India, in recent years, the country has come under increased risk of becoming diplomatically isolated due to the size of its economy and its emissions (now the world's third largest emitter) (Guardian 2014), despite its very low emissions per capita (ranked 147th among all countries in the world) (Guardian 2013).[3] In fact, India's per capita emissions are less than a third of China's. India walks a very interesting line in the climate negotiations. On the one hand, it has attempted to bolster itself as a major world leader with aspirations for a seat on the UN Security Council and a greater role in international financial institutions like the World Bank and the International Monetary Fund. On the other hand, it is a very poor country ranked 143rd in the Human Development Index, with 300 million people without access to electricity, that must appease national interests for meeting basic development needs, which many argue would be compromised by any limits on its emissions.

In the lead-up to the Paris negotiations, India often fought against being subject to limits on its growth and strongly advocated for "differentiation" in terms of responsibility for action. Coming into Paris, India put forward a more ambitious INDC than some expected.[4] However, unlike China, India had been unwilling to promise to peak its emissions in the future. India's pledge also came with a price tag of $2.5 trillion and a call to the international community to support its clean energy program and to help it to adapt to climate change between 2015 and 2030, in addition to seeking finance from the private sector (Sinha 2015). While its target for solar power is ambitious, it simultaneously set a target for coal production of 1.5 billion metric tons by 2020. Thus, India seems largely unwilling to commit to not developing its huge coal reserves, unless it is compensated for its behavior. In this sense, India may find itself in direct competition for climate finance resources with much smaller peripheral nations.

Additionally, unlike the LDCs and AOSIS, despite its high vulnerability to climate impacts, India has been reluctant to commit to a 1.5 degree Celsius limit on temperature change, playing a part in a group called the like-minded developing countries, and it has largely resisted movements toward reporting and transparency. In the Paris negotiations, in addition to continuing to demand differentiation between actor-groups in the agreement, India (like China) came under fire for standing against a rigorous five-year review of INDCs, supported by countries in the periphery. While India will increasingly represent a major global economy, political force, and net polluter, it is still comparable in many ways in terms of overall poverty, vulnerability, and emissions per capita to many LDCs. Still, it opposed several key positions of the LDC group.

A core practical concern is India's extreme vulnerability to climate change impacts, such as its dependence on glacier-fed water supplies from the Himalayas, its vast populations on semiarid lands with scarce irrigation, and its dense population in the coastal belt vulnerable to sea-level rise and intensifying monsoons. In monetary terms, the issue is clearly salient: the Indian government claims that 3 percent of its GDP is already being spent on adaptation to climate change impacts, and it will need $206 billion to cover related costs for the period of 2015–2030 (India 2015).

Overall, in the post-Paris period there is no longer a unifying position between the semi-peripheral states in the G-77 of maintaining the North–South divide enshrined in the Bali Agreements. In this context, the often competing and complex identities of China and India, along with those of other countries that occupy the middle of the global division of labor, may lead to increasing tensions within the G-77. This will likely take the form of an inability within the coalition to agree on ideas of equity, responsibility, differentiation, and accountability for climate action within the global order in the coming years.

Tensions Within the Periphery

Finally, there are important emerging tensions that may serve as wedges to solidarity within the global periphery in the negotiations and which make the ecologically unequal exchange discourse more difficult to maintain. First, there are potential tensions among peripheral states in the post-Paris context in terms of the extent to which actors maintain radical class-based demands concerning differentiation and compensation, or rather, embrace more fluid, pragmatic, or reformist ideas in UN climate politics (see, e.g. discussion of the AILAC coalition above). Although the 80 nations in AOSIS and the LDCs are highly vulnerable to climate change, the postures and positions of the individual countries often differ substantially across the groups. The particular states take varying stances on whether to challenge the positions of OPEC and BASIC, for example, often depending on the particular areas of conflict and who is chairing the groups. For example, the island states Tuvalu and the Maldives have gained attention for being far more ambitious and aggressive in their stances than have Saint Lucia or Samoa. Like the economically more powerful states in BASIC and OPEC, many LDC countries also pursue both bilateral and minilateral diplomacy with single countries or smaller groups to promote their individual and group interests (Khan 2013).

One issue where there has been some contention among peripheral states has been on their approach to the issue of loss and damage, which includes some form of help for those climate impacts that cannot be readily adapted to. Having been raised since the early 1990s by AOSIS, this issue first found great unity in the 2012 and 2013 Doha and Warsaw negotiations when peripheral states came together in coalitions (including the LDCs, AOSIS, the African Group, and the Central American Integration System, as well as the broader G-77), arguing for the establishment of a distinct loss and damage mechanism in the Convention. However, the specifics of this mechanism have been more controversial. While some actors, and particularly some states in AOSIS, have continually demanded that compensation and liability be a cornerstone of demands related to the loss and damage framework, other states have viewed this demand as either polarizing or unrealistic, and instead have focused on other less radical aspects of the program such as data management, research, and climate refugee legality. For example, as a concession to developed states such as the United States, the LDCs agreed to language in the Paris decision text for loss and damage that explicitly excluded compensation or liability of developing countries (Vanhala and Hestbaek 2016). However, countries like Tuvalu which are particularly threatened by rising seas and other catastrophic climate disasters are unlikely to give up the fight for compensation. This difference may serve as a wedge between peripheral states in terms differential exposure to loss and damage events, and differing positions on the issue in future negotiations.

Second, competition over scarce adaptation and other climate finance resources has been a wedge between peripheral states in recent years (Ciplet, Roberts, and Khan 2013b), and may intensify. The figures are stark: over 90 countries and their people have contributed an almost negligible amount to the problem of climate change, but they are already being hit first and hardest by the impacts, and they face these disasters with the least capacity to adapt to the changes (Kasperson and Kasperson 2001; Intergovernmental Panel on Climate Change 2007, 2013; Roberts and Parks 2006). As Desmond Tutu put it in 2008, a system of "adaptation apartheid" is already developing in the form of increasing investments in protections against climate-related disasters in industrial countries, while efforts in the most vulnerable countries have always been grossly underfunded (Tutu 2008).

The $30 billion in finance that wealthy states promised in Copenhagen for developing countries during the 2010–2012 period were not delivered as promised. In addition, there is limited evidence that a scale-up to the promised $100 billion a year by 2020 is taking place (Ciplet et al. 2015). There are many measures and much debate on both the supply and demand for adaptation aid (e.g. AdaptationWatch 2015; Ciplet et al. 2011; Oxfam America 2012), but the United Nations Environment Programme (2014) estimates that by 2025 or 2030 an estimated $150 billion of funding is needed to support adaptation to climate change in developing countries but current amounts of truly new public funds are still probably below $10 billion a year (Oxfam International 2015).

If such support is not dramatically scaled up, with new and substantial commitments during the next rounds of negotiations, states that are disproportionately vulnerable to climatic instability are likely to become more vocal in their demands for compensation, including for climate impacts that cannot be readily adapted to, such as rising seas. Importantly, the periphery may witness intensified infighting over designations of vulnerability in order to access the scarce existing public funds (Ciplet et al. 2015). This infighting could extend to the broader G-77 as conditions worsen and funds remain scarce.

While this may undermine efforts at collective organizing by peripheral states to address ecologically unequal exchange and remuneration for climate debt, this fragmentation is not inevitable. Notably, a coalition called the Climate Vulnerable Forum emerged as part of a conference in the Maldives in 2009. The Climate Vulnerable Forum was a key actor in Paris working to ensure that the 1.5 degree Celsius temperature target was included in relevant agreements. It is possible that the Climate

Vulnerable Forum can be a vehicle for vulnerable states to maintain strong collective demands across diverse negotiating blocs for climate finance demands.

Conclusion: Insights for Ecologically Unequal Exchange Theory

Existing conceptions of ecologically unequal exchange have provided very limited understanding of the distinct and nuanced political dynamics which shape how environmental inequality is governed globally. This has major implications for our understanding of the reproduction of inequality at a global scale, suggesting that the distribution of goods and bads in the global system is the result of not merely trade relations or military domination, but also interactions in the political realm of multilateral institutions. The case of the contemporary UNFCCC points to three main insights for a theory of ecologically unequal exchange governance and resistance.

First, the analysis suggests that ecologically unequal exchange theory must better consider the role of the semi-periphery in reproducing ecologically unequal governance forms. Existing scholarship has almost completely neglected the strategies employed by semi-peripheral states to maintain or enhance their relative ecological privilege in global environmental governance. The implicit underlying assumption has long been that the North is solely to blame for governance structures that support ecologically unequal exchange. However, in contemporary UNFCCC politics, semi-peripheral states have played a pivotal role in undermining robust mitigation efforts—particularly measures that would place limits on their own development aspirations. They have accomplished this by first dominating the G-77 bloc's positions and then later building alliances outside of the G-77 coalition to shirk their own responsibility to mitigate their emissions. By transporting carbon pollutants across national borders and becoming top-ranked nations in inflicting climate instability on the poorest and most vulnerable countries, these semi-peripheral actors are in effect creating a new ecologically unequal exchange. They have buffered themselves against resistance to their continued emissions by supporting the demands of peripheral states on issues such as climate finance, loss and damage against the global North, and (reluctantly) a target of 1.5 degree Celsius for maximum global average temperature change. A common characteristic of the BASIC group is that each nation is a regional power at risk of alienating many neighbors as it attempts to reach the world stage as a global leader (on Brazil, see Edwards and Roberts 2015). One could argue that their actions in the area of climate politics suggest that each is diminishingly concerned about alienating their regional neighbors and the rest of the G-77.

However, we have shown that due to the highly diverse economic and environmental positions among semi-peripheral states, ecologically unequal exchange theory should also be cognizant of the ways in which the semi-periphery, and its defined interests in regime politics, is not monolithic. To be sure, thus far, developing states as a group have committed through their climate plans to more emissions reductions during the 2020–2030 period than that of wealthy states, despite their significantly lower historical responsibility and ability to respond to the problem (Oxfam International 2015). But there are actors that are doing considerably more and less of their "fair share" to address the problem, as well as those that will be more or less vulnerable to the immediate consequences of warming climate.

Second, increasing fragmentation of defined interests between peripheral and semi-peripheral states may produce distinct challenges for peripheral states to resist governance forms which intensify ecologically unequal exchange. In this case, the changing landscape of major emitters in the global South—including countries like China, India, Brazil, Mexico, Turkey, and South Korea—has made it increasingly difficult for peripheral states to simply go along with the conventional wisdom that the North is solely or primarily responsible for taking action on climate change. We have discussed how in contemporary UNFCCC politics, peripheral state coalitions such as AOSIS, AILAC, and the LDCs have called for proposals that challenge a North–South binary for mitigation responsibility. At times, they have also formed alliances that cut across the North–South divide, such as "the Axis of Ambition" coalitions they formed with the EU in the negotiations in 2011 and 2015.

There is the distinct possibility that the main discursive underpinnings of demands for remuneration of the "climate debt" owed by the global North to the global South will have to be adapted to the changing emissions context. The post-Paris institutional conjuncture requires that all states take mitigation and adaptation action. This change opens new discursive opportunities to pressure not only wealthy states on the adequacy of their actions but also major rising polluters in the global South. It seems likely that as the poorest and most vulnerable states will experience increasingly intense climate disasters which are not of their own making, including the disappearance of whole territories under rising seas, that demands for compensation for climate debt will extend to other major polluters in the South, particularly if these states are unable or unwilling to commit to ambitious mitigation action or to fulfill their pledges within their INDCs. These tensions will also be amplified if there are not robust measures of accountability and transparency to ensure that actions outlined in INDCs are being fulfilled in practice—an issue upon which there is no agreement within the G-77.

However, such efforts of resistance from the global periphery come with major risks. There are also compelling reasons for why, even in this changing ecological and political context, peripheral states may be unwilling to take a stand against their larger and more economically developed allies within the G-77. Most notably, peripheral states would likely find less leverage in the negotiations against wealthy states on key issues such as climate finance and loss and damage without the support of their more economically, politically, and militarily strong semi-peripheral allies in the G-77 coalition. The increasingly strong financial aid and investment ties between China and states throughout Africa and Latin America also make public betrayal of conventional South–South ideals of solidarity potentially costly and present structural obstacles to resistance (Ciplet et al. 2015).

A more complex reality of the perpetrators of ecologically unequal exchange also has the potential to create further divisions among peripheral states, weakening their collective leverage for unified demands. Competition for scarce adaptation, mitigation, and loss and damage resources in a warming world may also lead to increased infighting among those most in need of support. India's own call for international support of $2.5 trillion for its INDC may further intensify tension between this rising semi-peripheral state, and between coalitions such as the like-minded developing countries and BASIC on the one hand, and the LDCs and AOSIS on the other. Such tensions may be offset by efforts for collective demands to address vulnerability, in groups such as the Climate Vulnerable Forum.

Given these challenges facing peripheral states, we pose a third challenge for ecologically unequal exchange theory going forward: it is important to analyze the ways in which unequal ecological exchange as mobilized as a collective action frame diverges from or conforms with the real-world distribution of environmental goods and bads in the world system. Specifically, in this context, even in a highly fragmented and increasingly multi-polar world system in which the biggest growth in climate pollution is in the global South, a simple North–South axis of political organization and identity may maintain utility and dominance for many global South state actors and coalitions. In the immediate term, peripheral states, given their structural and political weaknesses, may continue to make calculated decisions to play nice with their big friends in the South. This stance may be taken despite the risks that inadequate mitigation action by semi-peripheral countries poses to climate vulnerability in the periphery.

It may be in the nature of capitalism to accelerate fragmentation in the conditions of nations and to create unequal costs and benefits through broad systems of unequal exchange, both economic and ecological. But it seems that geopolitically, there will be times when nations choose to simplify their solidarity groups along North–South lines, and other times when they do not. This alignment may be the only one that functions to allow effective struggles for redistribution of economic benefits from those at the top of the global hierarchy. However, in the case of the unequal distribution of impacts of carbon pollution, such alignments may be highly contrary to the interests of those at the very bottom of the distribution, the most vulnerable.

In sum, we have argued that tensions within the G-77 coalition in the UN climate negotiations will largely influence the forms that struggles against ecological inequality in the world system take in the post-Paris period. These tensions point to the need for ecologically unequal exchange scholarship to move beyond primarily documenting the problem of unequal global material flows, to nuanced empirical exploration of the shifting political dimensions within the world system and specific governance contexts that shape opportunities for transformation. Such analysis that does take the real-world politics of resistance to ecologically unequal exchange seriously, and the politics of global climate justice in particular, should carefully consider the complex relationships and points of fragmentation within the strategic organization and identity of the global South.

Notes

1. We describe the development of the zones of world systems theory (periphery, semi-periphery, and core) below, and acknowledge that sometimes countries move in and out of these categories as they move up and down through the hierarchy of nations. While these are functional groups of nations, we also see these as a continuum from the most powerful and wealthy to the least developed countries. We refer to certain nations in each zone but do not base these categorizations on current empirical data; rather we utilize earlier world-system theory conceptualizations such as that of Terlouw (1993) (see Roberts and Grimes 2002).
2. This participant observation has included working directly with numerous policy NGOs and civil society networks and state delegations, particularly the least developed countries negotiating group. Observational data was collected during network meetings, side events, press conferences, demonstrations, and policy interventions. The analysis is also informed by more than 100 informal interviews and analysis of UNFCCC negotiations and related policy documents.
3. Ranking in per capita emission assessed using 2010 data.
4. India pledged to reduce its emissions intensity per unit of GDP 33 percent to 25 percent from 2005 levels by 2030, to receive about 40 percent of its power from non-fossil sources by 2030, and to enhance afforestation. It also announced a target to develop 100 GW of solar power capacity by 2022, launched a solar power alliance to increase solar production in the developing world, and has implemented a per ton tax on coal that is a direct subsidy to renewables.

References

AdaptationWatch. 2015. *Toward Mutual Accountability: The 2015 Adaptation Finance Transparency Gap Report*. Available at https://unfccc.int/files/cooperation_and_support/financial_mechanism/standing_committee/application/pdf/aw_report_24_11_15_lo_res.pdf.

Alessi, Christopher. 2012. "Expanding China-Africa Oil Ties." *Council on Foreign Relations*, February 8. Available at http://abdas.org/?p=1875.

Arrighi, Giovanni and Jessica Drangel. 1986. "The Stratification of the World-Economy: An Exploration of the Semiperipheral Zone." *Review (Fernand Braudel Center)* 10(1):9–74.

Bond, Patrick. 2010. "Climate Debt Owed to Africa: What to Demand and How to Collect?" *African Journal for Science, Technology, Innovation and Development* 2(1):83–113.

———. 2012. *Politics of Climate Justice: Paralysis Above, Movement Below*. Cape Town, South Africa: University of Kwa Zulu Natal Press.

Botzen, Wouter J. W., John M. Gowdy, and Jeroen C. J. M. van den Bergh. 2008. "Cumulative CO2 Emissions: Shifting International Responsibilities for Climate Debt." *Climate Policy* 8(6):569–576.

Bunker, Stephen G. 1985. *Underdeveloping the Amazon: Extraction, Unequal Exchange, and the Failure of the Modern State*. Chicago, IL: University of Chicago Press.

Burns, Thomas J., Edward L. Kick, and Byron L. Davis. 2003. "Theorizing and Rethinking Linkages between the Natural Environment and the Modern World-System: Deforestation in the Late 20th Century." *Journal of World-Systems Research* 9(2):357–390.

Burns, Thomas J., Byron Davis, and Edward L. Kick. 1997. "Position in the World-System and National Emissions of Greenhouse Gases." *Journal of World-Systems Research* 3(3):432–466.

Cardoso, Fernando Henrique and Enzo Faletto. 1979. *Dependency and Development in Latin America*. Berkeley, CA: University of California Press.

Center for Global Development. 2015. "Developing Countries are Responsible for 63 Percent of Emissions." August 18.

Chase-Dunn, Christopher K. 1998. *Global Formation: Structures of the World-Economy*. Lanham, MD: Rowman & Littlefield.

Chatterton, Paul, David Featherstone, and Paul Routledge. 2013. "Articulating Climate Justice in Copenhagen: Antagonism, the Commons, and Solidarity." *Antipode* 45(3):602–620.

Chayes, Abraham and Charlotte J. Kim. 1998. "China and the United Nations Framework Convention on Climate Change." Pp. 503–540 in *Energizing China: Reconciling Environmental Protection and Economic Growth*, edited by Michael B. McElroy, Chris P Nielsen, and P. Lydon. Cambridge, MA: Harvard University Press.

Ciplet, David. 2014. "Contesting Climate Injustice: Transnational Advocacy Network Struggles for Rights in UN Climate Politics." *Global Environmental Politics* 14(4):75–96.

———. 2015. "Rethinking Cooperation: Inequality and Consent in International Climate Change Politics." *Global Governance* 21(2):247–274.

Ciplet, David, J. Timmons Roberts, Martin Stadelmann, Saleemul Huq, and Achala Chandani. 2011. *Scoring Fast Start Climate Finance: Leaders and Laggards in Transparency*. International Institute for Environment and Development. Available at http://pubs.iied.org/17100IIED/?s=IIEDBRIEF.

Ciplet, David, J. Timmons Roberts, Pa Ousman, Achala Abeysinghe, Alexis Durand, Daniel Kopin, Olivia Santiago, Keith Madden, and Sophie Purdom. 2013a. *A Burden to Share? Addressing Unequal Climate Impacts in the Least Developed Countries*. International Institute for Environment and Development. Available at http://pubs.iied.org/pdfs/17181IIED.pdf.

Ciplet, David, J. Timmons Roberts, and Mizan Khan. 2013b. "The Politics of International Climate Adaptation Funding: Justice and Divisions in the Greenhouse." *Global Environmental Politics* 13 (1):49–68.

———. 2015. *Power in a Warming World: The New Global Politics of Climate Change and the Remaking of Environmental Inequality*. Cambridge, MA: MIT Press.

Downey, Liam. 2015. *Inequality, Democracy, and the Environment*. New York: New York University Press.

Doyle, A. 2015. "China to Surpass US as Top Cause of Modern Global Warming." *Reuters*, April 13.

Earth Negotiations Bulletin. 1995. "Working Group 1." December 11. Available at http://www.iisd.ca/vol12/1211011e.html.

Economy, E. 1997. "Chinese Policy-Making and Global Climate Change: Two-Front Diplomacy and the Internationality Community." Pp. 19–41 in *The Internationalization of Environmental Protection*, edited by M. Schreurs and E. Economy. Cambridge: Cambridge University Press.

Edwards, Guy and J. Timmons Roberts. 2015. *A Fragmented Continent: Latin America and the Global Politics of Climate Change*. Cambridge, MA: MIT Press.

Energy Information Administration. 2013. *International Energy Outlook 2013*. Available at https://www.eia.gov/outlooks/ieo/pdf/0484(2013).pdf.

Escobar, Arturo. 1998. "Whose Knowledge, Whose Nature? Biodiversity, Conservation, and the Political Ecology of Social Movements." *Journal of Political Ecology* 5(1):53–82.

Frey, R. Scott. 2015. "Breaking Ships in the World-System: An Analysis of Two Ship Breaking Capitals, Alang-Sosiya, India and Chittagong, Bangladesh." *Journal of World-Systems Research* 21(1):25–49.

Friedman, Lisa. 2013a. "China's Friendship with Costa Rica May Conflict with Its Green Goals." *ClimateWire*, June 20.

———. 2013b. "New Alliance Attempts to Bridge North-South Gap in U.N. Climate Talks." *ClimateWire*, January 22.

———. 2015. "Turkey: Can a Country Planning 80 Coal-Fired Power Plants Get Serious about Climate Change?" *ClimateWire*, October 9.

Guardian. 2013. "World Carbon Dioxide Emissions by Country." *The Guardian*, July 15.

———. 2014. "World Carbon Dioxide Emissions Data by Country: China Speeds Ahead of the Rest." *The Guardian*, January 31.

Gupta, Joyeeta. 1997. *The Climate Change Convention and Developing Countries: From Conflict to Consensus?* Vancouver, Canada: Springer.

Hart, Melanie, Pete Ogden, and Greg Dotson. 2015. "U.S.-China Climate Announcements Lay Foundation for Success in Paris." Center for American Progress. https://www.americanprogress.org/issues/security/news/2015/09/25/122098/u-s-china-climate-announcements-lay-foundation-for-success-in-paris/

Hecht, Susanna, Anthony Anderson, and Peter May. 1988. "The Subsidy from Nature: Shifting Cultivation, Successional Palm Forests, and Rural Development." *Human Organization* 47(1):25–35.

Heggelund, G. 2007. "China's Climate Change Policy: Domestic and International Developments." *Asian Perspective* 31:155–191.

Hornborg, Alf. 2001. *The Power of the Machine: Global Inequalities of Economy, Technology, and Environment*. Lanham, MD: Rowman Altamira.

Hurrell, Andrew and Sandeep Sengupta. 2012. "Emerging Powers, North–South Relations and Global Climate Politics." *International Affairs* 88(3):463–484.

India. 2015. "India's Intended Nationally Determined Contribution: Working toward Climate Justice." Available at http://www4.unfccc.int/submissions/INDC/Published%20Documents/India/1/INDIA%20INDC%20TO%20UNFCCC.pdf.

Intergovernmental Panel on Climate Change. 2007. *Synthesis Report: Climate Change 2007: Impacts, Adaptation and Vulnerability*, edited by O. F. Canziani,

M. L. Parry, J. P. Palutikof, P. J. van der Linden, and C. E. Hanson. Cambridge: Cambridge University Press.

———. 2013. *Climate Change 2013: The Physical Science Basis*, edited by T. F. Stocker, D. Qin, G. K. Plattner, M. Tignor, S. K. Allen, J. Boschung, A. Nauels, Y. Xia, B. Bex, and B. M. Midgley. Cambridge: Cambridge University Press.

Jorgenson, Andrew K. 2004. "Uneven Processes and Environmental Degradation in the World-Economy." *Human Ecology Review* 11(2):103–117.

Jorgenson, Andrew K. and Brett Clark. 2009. "The Economy, Military, and Ecologically Unequal Exchange Relationships in Comparative Perspective: A Panel Study of the Ecological Footprints of Nations, 1975–2000." *Social Problems* 56(4):621–646.

Kasperson, Roger E. and Jeanne X. Kasperson. 2001. *Climate Change, Vulnerability, and Social Justice*. Stockholm: Stockholm Environment Institute.

Katz, Ian. 2014. "China's Treasury Holdings Climate to Record in Government Data." *Bloomberg Business*, January 15.

Khan, Mizan. 2013. *Towards a Binding Climate Change Adaptation Regime: A Proposed Framework*. London: Routledge.

Khor, Martin. 2015. "Third World Network Bulletin." *Third World Network*, November 10.

Klein, Naomi. 2010. "Paying Our Climate Debt." Pp. 71–76 in *Climate Change—Who's Carrying the Burden?: The Chilly Climate of the Global Environmental Dilemma*, edited by L. Anders Sandberg and Tor Sandberg. Ottawa, Canada: Canadian Center for Policy Alternatives.

Krasner, Stephen D., 1985. *Structural Conflict: The Third World Against Global Liberalism*. Berkeley, CA: University of California Press.

Mathiesen, Karl and Fiona Harvey. 2015. "Climate Coalition Breaks Cover in Paris to Push for Binding and Ambitious Deal." *The Guardian*, December 8.

Mauldin, William. 2015. "China's Xi Wants More Funds from Rich Nations in Climate Deal." *Wall Street Journal*, November 30.

McGrath, M. 2014. "China's Per Capita Carbon Emissions Overtake EU's." *BBC*, September 21.

Méndez, Rafael. 2012. "La Tercera vía Latinoamericana en la Negociación del Clima." *El País*, December 5. Available at http://sociedad.elpais.com/sociedad/2012/12/05/actualidad/1354699047_259945.html.

Mol, Arthur P. and Gert Spaargaren. 2000. "Ecological Modernization Theory in Debate: A Review." *Environmental Politics* 9(1):17–49.

Najam, Adil. 2005. "Why Environmental Politics Looks Different from the South." Pp. 111–126 in *Handbook of Global Environmental Politics*, edited by Peter Dauvergne. Cheltenham, UK: Edward Elgar.

Nepal. 2015. "Submission by Nepal on Behalf of the LDCs Group on the ADP Workstream 1." Available at http://www.1o5c.org/wp-content/uploads/2015/09/LDC.pdf.

O'Brien, Karen L. and Robin M. Leichenko. 2000. "Double Exposure: Assessing the Impacts of Climate Change within the Context of Economic Globalization." *Global Environmental Change* 10(3):221–232.

Okereke, Chukwumerije. 2006. "Global Environmental Sustainability: Intragenerational Equity and Conceptions of Justice in Multilateral Environmental Regimes." *Geoforum* 37(5):725–738.

Oxfam America. 2012. *The Climate Fiscal Cliff: An Evaluation of Fast Start Finance and Lessons for the Future*. Available at https://www.oxfam.org/sites/www.oxfam.org/files/oxfam-media-advisory-climate-fiscal-cliff-doha-25nov2012.pdf.

Oxfam International. 2015. *Fair Shares: A Civil Society Equity Review of INDCs*. Available at https://www.oxfam.org/sites/www.oxfam.org/files/file_attachments/ib-civil-society-review-climate-indcs-191015-en_2.pdf.

Pettit, J. 2004. "Climate Justice: A New Social Movement for Atmospheric Rights." *IDS Bulletin* 35(3):102–106.

Pickering, Jonathan and Christian Barry. 2012. "On the Concept of Climate Debt: Its Moral and Political Value." *Critical Review of International Social and Political Philosophy* 15(5):667–685.

Prebisch, Raul. 1950. *The Economic Development of Latin America and Its Principal Problems*. New York: United Nations.

Rice, James. 2007. "Ecological Unequal Exchange: International Trade and Uneven Utilization of Environmental Space in the World System." *Social Forces* 85(3):1369–1392.

Roberts, David. 2011. "The Top Five Takeaways from the Durban Climate Talks." *Grist*, December 11. Available at http://grist.org/climate-change/2011-12-12-the-top-five-takeaways-from-the-durban-climate-talks/.

Roberts, J. Timmons and Peter E. Grimes. 2002. "World-System Theory and the Environment: Toward a New Synthesis." Pp. 167–196 in *Sociological Theory and the Environment: Classical Foundations, Contemporary Insights*, edited by Frederick H. Buttel, August Gijswijt, Peter Dickens, and Riley E. Dunlap. Lanham, MD: Rowman & Littlefield.

Roberts, J. Timmons and Bradley C. Parks. 2006. *A Climate of Injustice: Global Inequality, North-South Politics, and Climate Policy*. Cambridge, MA: MIT Press.

———. 2007. "Fueling Injustice: Globalization, Ecologically Unequal Exchange and Climate Change." *Globalizations* 4(2):193–210.

———. 2009. "Ecologically Unequal Exchange, Ecological Debt, and Climate Justice: The History and Implications of Three Related Ideas for a New Social Movement." *International Journal of Comparative Sociology* 50(3–4):385–409.

Rothstein, R. L. 2015. *Global Bargaining: UNCTAD and the Quest for a New International Economic Order*. Princeton, NJ: Princeton University Press.

Schiavenza, Matt. 2014. "China Economy Surpasses US in Purchasing Power, But Americans Don't Need to Worry." *International Business Times*, October 8.

Schroeder, Heike. 2010. "Agency in International Climate Negotiations: The Case of Indigenous Peoples and Avoided Deforestation." *International Environmental Agreements: Politics, Law and Economics* 10(4):317–332.

Shandra, John M., Christopher Leckband, Laura A. McKinney, and Bruce London. 2009. "Ecologically Unequal Exchange, World Polity, and Biodiversity Loss: A Cross-National Analysis of Threatened Mammals." *International Journal of Comparative Sociology* 50(3–4):285–310.

Shiva, Vandana. 1996. "Agricultural Biodiversity, Intellectual Property Rights and Farmers' Rights." *Economic and Political Weekly* 31(25):1621–1631.

Singer, H. W. 1950. "The Distribution of Gains between Investing and Borrowing Countries." *American Economic Review* 40(2):473–485.

Sinha, Amitabh. 2015. "Here Are India's INDC Objectives and How Much It Will Cost." *The Indian Express*, October 2.

Smith, Christopher. 2010. "Bali Firewall and Member States' Future Obligations within the Climate Change Regime." *The Law Environment and Development Journal* 6:284.

Terlouw, C. P. 1993. "The Elusive Semiperiphery: A Critical Examination of the Concept of Semiperiphery." *International Journal of Comparative Sociology* 34(1):87–102.

Terry, Geraldine. 2009. "No Climate Justice without Gender Justice: An Overview of the Issues." *Gender & Development* 17(1):5–18.

Tutu, Desmond. 2008. "We Do Not Need Climate Change Apartheid in Adaptation." Human Development Report 2007/2008. New York: United Nations Development Programme.

United Nations Environment Programme. 2014. *The Adaptation Gap Report: A Preliminary Assessment*. Nairobi, Kenya: United Nations Environment Programme.

Vanhala, Lisa and Cecilie Hestbaek. 2016. "Framing Climate Change Loss and Damage in the UNFCCC Negotiations." *Global Environmental Politics* 16(4):111–129.

Vidal, John and David Adam. 2007. "China Overtakes US as World's Biggest CO2 Emitter." *The Guardian*, June 19.

Vihma, Antto. 2010. "Elephant in the Room: The New G77 and China Dynamics in Climate Talks." Briefing Paper 6, The Finnish Institute of International Affairs.

Vihma, Antto, Yacob Mulugetta, and Sylvia Karlsson-Vinkhuyzen. 2011. "Negotiating Solidarity? The G77 Through the Prism of Climate Change Negotiations." *Global Change, Peace & Security* 23(3):315–334.

Wallerstein, Immanuel. 1979. *The Capitalist World-Economy*. Cambridge, MA: Cambridge University Press.

———. 2011. *The Modern World-System I: Capitalist Agriculture and the Origins of the European World-Economy in the Sixteenth Century*. Berkeley, CA: University of California Press.

12

Epilogue: The Wider View

Harry F. Dahms and R. Scott Frey

As the essays included in this volume demonstrate in a broad range of ways, with regard to the scope and impact of resource extraction and exploitation as well as waste exports, ecologically unequal exchange (EUE) denotes a seemingly inexorable process whose impetus is both global and planetary in nature and in scale. It is global in that for some time, especially since Immanuel Wallerstein's (1974) world-systems analysis, it has been neither possible nor justifiable to try to explain conditions in individual societies and nation-states independently of global processes. Evidently, this applies both to the Global South and to the Global North, though in different ways. The exploitative regime that has structured and organized the condition of humankind in the world-system since the nineteenth century is so integral to the realities of our species and of the planet we inhabit that the social sciences perpetually are not only in danger of proceeding on the basis of increasingly unrealistic

We thank Paul Gellert for helpful comments.

H. F. Dahms (✉) • R. S. Frey
University of Tennessee, Knoxville, TN, USA
e-mail: hdahms@utk.edu; rfrey2@utk.edu

R. S. Frey et al. (eds.), *Ecologically Unequal Exchange*,
https://doi.org/10.1007/978-3-319-89740-0_12

assumptions, but they are increasingly likely to do so. Typically, the study of the social, political, and economic structures and systems implicitly assumes that it is possible to identify and delineate general patterns from within the context of any society, while neglecting the specificity of prevailing conditions at the national level, and without paying close attention to how the tentacles of the global economic system—the world-system—both reach far into and shape the purportedly autonomous and more or less peculiar configurations that are at work and at play in particular countries.

Both societies in the Global South and in the Global North would look very different today, had the EUE process not been in place for centuries; they most certainly would not be the societies that they are now, in a multiplicity of ways. In all likelihood, societies in the Global South would be fraught with pathologies and contradictions to a lesser extent, or with pathologies and contradictions of other types along the lines of the cultural traditions and normative frameworks in place in these societies, respectively (as opposed to the pathologies characteristic of the more "advanced" societies of the Global North). On the one hand, greater access to and control over their own resources presumably would alleviate economic strain and deprivation resulting from EUE. On the other hand, these societies still would struggle with internal tensions and contradictions and suffer from social conflicts, inequalities, and injustices, along with other types of burdens that come with progress under conditions of modernization—such as too much or too rapid progress, or too little, or not the right kind, depending on the vantage point applied and the segment of a given population focused on or examined. Still, we can only speculate about the extent and the ways in which societies in the Global South would have evolved, especially since the end of World War II, without the global regime of EUE having been in place, as it continued to intensify during the second half of the twentieth century, and even more so since the beginning of the current century. Presumably, though, there would have been more and greater opportunities in those societies to develop strategies and policies designed for and attuned to confronting challenges that were specific to and cognizant of the particulars "on the ground."

At the same time, absent proliferating opportunities to extract resources and wealth from and offshore anti-wealth to other places, the societies of the Global North inevitably would be fraught today to a discernibly greater extent by the contradictions that characterize modern industrialized capitalist societies with more or less nominally democratic political systems. With less access to natural resources and environmental space from the Global South, and thus with less economic wealth, it would be much more difficult for elites in the Global North who benefit most from domestic structures of inequality and systems of economic and political control and power to distract from and plaster over inevitable conflict lines and struggles over resources and access to privileges "at home." This also applies to the ability of societies of the Global North to manage consequences resulting from international competition among societies at similar stages of what used to be referred to as development, and to contain the destructive potential located in such competition.

As has become evident in recent years (and even more so since the EUE conference was held at the University of Tennessee in fall of 2015), in the absence of sustained efforts to acknowledge and remedy the strained state of suspended animation within and between the Global South and the Global North, authoritarian nationalist responses to, and strategies to manage, proliferating national and international tensions, challenges and crises have become increasingly acceptable, if not appealing, to large segments of the population in many countries. Prime examples are Turkey, Poland, Hungary, the Philippines, the United States and—most recently, with the abolition of constitutionally required Presidential term limits—in China. Indeed, the evidence suggests that the retreat of democracy predicted during the last decade of the twentieth century is continuing, and gaining momentum.[1] The retreat and weakening of democratic processes and institutions do not bode well for efforts to reign in, redirect, or contain the process of EUE—quite the opposite.

The seemingly inexorable impetus of EUE also is evident at the planetary level. While in the *global* context, despite the myriad empirical fault lines and public policy challenges, the process of *globalization*—regardless of whether it is understood as "creative destruction" or the opening of markets both old and new—appears to point toward and be consistent with infinite opportunities for economic expansion, with Earth purportedly constituting an open system, the ideological character of related visions and notions is becoming more and more difficult to deny. Yet, when conceived of in terms of planetary categories and considerations, it is undeniable that our planet less and less resembles an open system, and is described more accurately along the lines of what Kenneth Boulding in the 1960s referred to as "Spaceship Earth."[2] De facto, the Global North has not only been exploiting the Global South by establishing and solidifying a system of structural exploitation; rather, the latter is indicative of a pattern of *consuming* the planet for the purpose of increasing economic profits, leaving destruction and the potential for proliferating catastrophes in its wake, and materially altering at least the outer layers of the planet. This pattern also is observable within the Global North, internally, especially so in recent months, as intensifying and increasingly successful efforts are underway in the United States to undercut, remove, or abolish environmental legislation and protections that have been in place for decades, but which were not particularly strong to begin with. What has been happening to the material world we inhabit on the planetary scale is happening within the confines of the nation-state along similar lines, and vice versa, thus maintaining a process of destruction whose force-field is so common and so prevalent—so "natural"—as to be taken for granted by most individuals across all societies.

The major issues discussed in this volume provide evidence for examining the logic underlying EUE in ways that point toward research agendas and projects which are directly concerned with increasingly problematic and disturbing processes and developments. Inevitably, gaps remain in the existing literature that must be addressed and filled, as the process of EUE continues to unfold. Yet even if these gaps were to be filled, it is apparent that researchers should not be too hopeful—in fact, not even hopeful—as far as expectations are concerned to the effect that more rigorous, pointed, and critical research will have a positive bearing on the underlying logic of EUE. Rather, if the evidence does suggest a link between the efficacy of research and the underlying logic, which is invisible and not directly observable, it may be inversely related to hopes and expectations that inform and inspire research agendas, projects, and strategies. The above-mentioned logic may well have the capacity to use new knowledge for its sake, rather than for the sake of humankind, than for the purposes for which researchers have been endeavoring to illuminate the process, and will continue to. How is this so?

To start with, there is compelling evidence for the logic underlying EUE being the preliminary end result of an ongoing, adaptive, and evolutionary process at the societal, global, and planetary levels that have been occurring under our feet, behind our backs, and above our heads. And, if indeed so, we must ask whether researchers are willing and able to recognize in research agendas relating to EUE the prospect of this logic being sufficiently cunning to solidify further the divide between core and periphery to reconfigure societies on both sides accordingly—under the radar, as it were—and whether researchers are able to be cognizant of resulting dilemmas with consistency. Paradoxically, if the logic is truly cunning, it may be capable of incorporating insights gained in social research about EUE in its operations, as they are quintessentially *non-human*: rather than us being in a position to guide and utilize it for our purposes, it has been performing a key role in structuring both our world and our experiences, and especially how we relate to others, in ways that are not just difficult to conceive, but which also are painful to face, in light of the justice and fairness-based norms and values that are supposed to inform and orient the actions of modern democratic men and women.

The question that must be posed, then, is whether this process of EUE, in its myriad manifestations, is consistent with and conducive to the kind of narratives that require of social scientists and social researchers a shift in focus, especially with regard to notions and conceptions of progress and the changing role of human agency. What most certainly is called for is a shift in rhetoric: it is no longer appropriate, and possibly not even justifiable, to posit that decision-makers, nation-states, and global actors have the capacity to halt or redirect the process, even though institutionally established, condoned, or required linguistic codes demand that the issues at hand are framed constructively and productively. The best we can do for the time being may well be acknowledging explicitly the likely futility of efforts to influence—to slow or to stop—the thrust and direction of EUE. At the very least, we would be well-advised to hypothesize that for the foreseeable future, "resistance [may be] futile," and our emphasis may have to be placed on "bearing witness" (i.e., tracking related processes and gather detailed information), until objective circumstances change, for example, as a consequence of imminent dangers that would threaten not just human beings—potentially a perfectly acceptable form of collateral damage, as far as the logic of EUE is concerned—but more importantly, the continued stability and functioning of the global economic system.[3]

The logic underlying EUE suggests that rather than moving toward reconciliation of society and nature, since the beginning of industrialization under the aegis of the spreading capitalist mode of production, the history of modern capitalist society from the outset was pointing toward nature being turned into a "function" of society—with society being a function of the logic of capital. From an Enlightenment perspective, human action and social activities in politics, culture, and the economy should be fueling ongoing transformations that are pointing toward circumstances on Earth in which shared norms and values are being actualized to an ever greater extent, over time. Yet, as classical social theorists and sociologists in the nineteenth and early twentieth century insisted, the precepts of the Enlightenment, while important and productive in many ways, were not consistent with the nature and logic of modern capitalist societies. Instead, while Enlightenment notions to the effect that human beings should be in charge of macro-social transformations are perfectly understandable and indeed desirable, they also are incongruous with what the founders of sociology taught us about modern society: that individuals are "functionaries of society," as Emile Durkheim put it, and that modern society is mediated by capital and corresponding modes of organization and technology, rather than an expression and the manifestation of human sociality.[4]

From the perspective of the early twenty-first century, trends appear to be pointing at an accelerating pace toward the formation of conditions that amount to an inversion of Enlightenment notions and expectations, along with a reversal of the philosophy of history that emerged from the latter: rather than moving toward a reconciliation of society and nature, there is mounting evidence suggesting that the trajectory of historical development in modern society is pointing toward nature being eradicated in favor of society, with society and the environment in which it continues to grow, becoming ever more artificial. According to the logic of EUE, nature is there to be consumed and replaced by more predictable, reliable, and manageable substitutes that guarantee that the logic of capital will persist forever.

To social scientists and social theorists, identifying and scrutinizing the logic of capital, and by implication, the logic underlying the process of EUE, has been a task of the first order, even though each of the social sciences and their routine concerns and practices must be understood as at least to a certain extent expressions, rather than critical analyses, of both logics, and transpositions of the latter onto the level of implicit presuppositions that inform an array of political, policy-related, research, ideological, religious and normative agendas and concerns, and which thus replicate and even conceal this logic.

With society in the modern age being synonymous with *capitalist* society—a society dominated by a particular type of economics and of "doing business"—the process of EUE may well be the harbinger of "artifice": a type of socio-economic organization, characterized by corresponding modes of politics and culture, in whose context "nature" is being transformed into constructs that undergird non-human economic logics which humans, in turn, misinterpret as expressions of their own nature, and of human sociality generally.[5]

Modern society thus appears as a type of socio-economic organization that is characterized by corresponding modes of politics and culture in whose context "nature" is being supplanted, to an ever increasing extent, by artificial substitutes. Paradoxically, with *artifice* potentially turning out to be the real vanishing point of modern society, social, political, and cultural forms will be replaced by modes of "human" and "social" interaction (and corresponding, new institutional structures) that further replicate, reinforce, amplify, and aggravate organizational, technological, and economic patterns at the heart of capitalism.

The possibility to avert this prospect may still exist; if it is to be avoided, however, equalizing ecological exchange around the globe would be a necessary precondition. It is conceivable that such equalizing would have the potential of being the necessary *first step* toward remedying the world-system's impetus toward maintaining and amplifying existing forms of inequality that the tools of nation-states no longer are capable of containing or even alleviating. Such equalizing may well be required for remedying the world-system's impetus toward maintaining and amplifying existing forms of inequality that nation-states do not have the capacity to control, if they ever did. Furthermore, such equalizing also would necessitate a reconstruction of the structure of individual identity, along with a radical reconceptualization of the reference frame for social analysis, for social research, and for sociology (see Dahms 2018). To confront global and planetary challenges constructively, we will need to play close attention to how the larger context in which we live our lives reverberates in our practices, social relations, and habits of perceiving the world.

Notes

1. See Kaplan (1997); also Kurlantzick (2013); Diamond and Plattner (2015); Diamond, Plattner, and Walker (2016); The Data Team (2018).
2. "The closed earth of the future requires economic principles which are somewhat different from those of the open earth of the past. For the sake of picturesqueness, I am tempted to call the open economy the "cowboy economy," the cowboy being symbolic of the illimitable plains and also associated with reckless, exploitative, romantic, and violent behavior, which is characteristic of open societies. The closed economy of the future might similarly be called the "spaceman" economy, in which the earth has become a single spaceship, without unlimited reservoirs of anything, either for extraction or for pollution, and in which, therefore, man must find his place in a cyclical ecological system which is capable of continuous reproduction of material form even though it cannot escape having inputs of energy. The difference between the two types of economy becomes most apparent in the attitude towards consumption. In the cowboy economy, consumption is regarded as a good thing and production likewise; and the success of the economy is measured by the amount of the throughput from the 'factors of production,' a part of which, at any rate, is extracted from the reservoirs of raw materials and noneconomic objects, and another part of which is output into the reservoirs of pollution. …By contrast, in the spaceman economy, throughput is by no means a desideratum, and is indeed to be regarded as something to be minimized rather than maximized. The essential measure of the success of the economy is not production and consumption at all, but the nature, extent, quality, and complexity of the total capital stock, including in this the state of the human bodies and minds included in the system. In the spaceman economy, what we are primarily concerned with is stock maintenance, and any technological change which results in the maintenance of a given total

stock with a lessened throughput (that is, less production and consumption) is clearly a gain. This idea that both production and consumption are bad things rather than good things is very strange to economists, who have been obsessed with the income-flow concepts to the exclusion, almost, of capital-stock concepts" (Boulding 1966:9–10). See also Boulding ([1973] 1980) and Spash (2013).
3. "The emergence of the possibility of a future, in which surplus production no longer must be based on the labor of an oppressed class, is, at the same time, the emergence of the possibility of a disastrous development in which the growing superfluity of labor is expressed as the growing superfluity of people" (Postone 2015:21).
4. On the concept of the "logic of capital," see Dahms (2015a, 2015b). Regarding the notion of "functionaries of society," see Durkheim ([1893] 1997:28).
5. Regarding the concept of artifice as a social-theoretical concept, see Dahms (2017).

References

Boulding, Kenneth E. 1966. "The Economics of the Coming Spaceship Earth." Pp. 3–14 in *Environmental Quality in a Growing Economy: Essays from the Sixth RFF Forum*, edited by Henry Jarrett. Baltimore, MD: Johns Hopkins University Press.

———. [1973] 1980. "Spaceship Earth Revisited." Pp. 264–266 in *Economics, Ecology, Ethics: Essays toward a Steady-State Economy*, edited by Herman E. Daly. New York: W. H. Freeman.

Dahms, Harry F. 2015a. "Which Capital, Which Marx? Basic Income between Mainstream Economics, Critical Theory, and the Logic of Capital." *Basic Income Studies* 10(1):115–140.

———. 2015b. "Toward a Critical Theory of Capital in the 21st Century: Thomas Piketty between Adam Smith and the Prospect of Apocalypse." *Critical Sociology* 41(2):359–374.

———. 2017. "Critical Theory in the Twenty-First Century: The Logic of Capital between Classical Social Theory, the Early Frankfurt School Critique of Political Economy, and the Prospect of Artifice." Pp. 47–74 in *The Social Ontology of Capitalism*, edited by Daniel Krier and Mark Worrell. New York: Palgrave Macmillan.

———. 2018. "Critical Theory, Radical Reform and Planetary Sociology: Between Impossibility and Inevitability." Pp. 152–168 in *Piketty, Inequality and 21st Century Capitalism*, edited by David A. Smith and Lauren Langman. Leiden, The Netherlands: Brill.

Diamond, Larry and Marc F. Plattner, editors. 2015. *Democracy in Decline?* Baltimore, MD: Johns Hopkins University Press.

Diamond, Larry, Marc F. Plattner, and Christopher Walker, editors. 2016. *Authoritarianism Goes Global. The Challenge to Democracy*. Baltimore, MD: Johns Hopkins University Press.

Durkheim, Emile. [1893] 1997. *The Division of Labor in Society*. New York: The Free Press.

Kaplan, Robert. 1997. "Was Democracy Just a Moment?" *The Atlantic Monthly*, December. Available at https://www.theatlantic.com/magazine/archive/1997/12/was-democracy-just-a-moment/306022/.

Kurlantzick, Joshua. 2013. *Democracy in Retreat: The Revolt of the Middle Class and the Worldwide Decline of Representative Government*. New Haven, CT: Yale University Press.

Postone, Moishe. 2015. "The Task of Critical Theory Today: Rethinking the Critique of Capitalism and Its Futures." Pp. 3–28 in *Globalization, Critique and Social Theory: Diagnoses and Challenges*, edited by Harry F. Dahms (Current Perspectives in Social Theory 33). Bingley, UK: Emerald.

Spash, Clive L. 2013. "The Ecological Economics of Boulding's Spaceship Earth." SRE-Discussion Papers 3919, WU Vienna University of Economics and Business. Available at http://www-sre.wu.ac.at/sre-disc/sre-disc-2013_02.pdf.

The Data Team. 2018. "Democracy Continues Its Disturbing Retreat." *The Economist*, January 31. Available at https://www.economist.com/blogs/graphicdetail/2018/01/daily-chart-21.

Wallerstein, Immanuel. 1974. *The Modern World-System I: Capitalist Agriculture and the Origins of the European World-Economy in the Sixteenth Century*. New York: Academic Press.

Index[1]

[1] Note: Page numbers followed by 'n' refer to notes.

R. S. Frey et al. (eds.), *Ecologically Unequal Exchange*,
https://doi.org/10.1007/978-3-319-89740-0

J

W

Y

Z

The manufacturer's authorised representative in the EU is Springer Nature Customer Service Centre GmbH, Europaplatz 3, 69115 Heidelberg, Germany. If you have any concerns regarding our products, please contact ProductSafety@springernature.com

Printed and bound by CPI Group (UK) Ltd, Croydon, CR0 4YY

30/09/2026

02225654-0003